Isabel Honorata de Souza Azevedo

Bioindikatoren, Ästuar, Foraminiferen, Mangroven

Isabel Honorata de Souza Azevedo

Bioindikatoren, Ästuar, Foraminiferen, Mangroven

Environmental assessment of estuaries and mangrove areas

Imprint

Any brand names and product names mentioned in this book are subject to trademark, brand or patent protection and are trademarks or registered trademarks of their respective holders. The use of brand names, product names, common names, trade names, product descriptions etc. even without a particular marking in this work is in no way to be construed to mean that such names may be regarded as unrestricted in respect of trademark and brand protection legislation and could thus be used by anyone.

Cover image: www.ingimage.com

This book is a translation from the original published under ISBN 978-613-9-62475-1.

Publisher:
Sciencia Scripts
is a trademark of
Dodo Books Indian Ocean Ltd. and OmniScriptum S.R.L publishing group

120 High Road, East Finchley, London, N2 9ED, United Kingdom
Str. Armeneasca 28/1, office 1, Chisinau MD-2012, Republic of Moldova, Europe
Printed at: see last page
ISBN: 978-620-7-74570-8

SUMMARY

ACKNOWLEDGMENTS

First of all, I would like to thank God for his protection and guardianship during all the stages of this work.

I would also like to thank my two mothers, Raquel Martins de Souza and Antônia Pereira de Almeida, who have always been there for me throughout my school and now academic life, collaborating, understanding, supporting me and worrying about me.

I would like to thank my friends Juliete Vidal and Maria Luiza Morely, for the many celebrations, for their trust, for understanding my absences and always supporting me.

I would like to thank my great friend, Wagner Magalhaes, who, even from a distance, never stopped worrying about me, for his support, consideration and patience in reading and correcting many of my works, as well as always being present in my life.

My special thanks go to Tais dos Santos Costa for the great moments of joy, relaxation, friendship and constant support.

I would like to thank my friends from the GEF group, Diogenes, Marcus, Matheus and Ruth, for the many relaxing moments I spent under the stereomicroscope, and Amanda and Vitor for their help sorting the samples.

I would like to thank the friends I made at POSPETRO, Ana Carina Matos, Andressa Nery, Nûria Mariana Campos and Daiane Oliveira, for all their help during the development of the work, for all their support and friendship; Narayana, Fabiany, Josana, Carine, Maria, Maria Luiza, Leila Maria, Leila Oliveira and Sheila for the many laughs they gave me during the most difficult stages.

I would like to thank Daniela Anunciaçao, "Dani", for her endless help with the statistical analysis, her corrections, her help with the oral presentations and her words of encouragement whenever I needed them.

To the LEPETRO technicians, Gisele, Sarah and Jorginho, for always being helpful and willing to help with the use of the chapel.

I would like to thank my advisors, Prof. Dr. Antônio Fernando, for trusting my work and giving me the opportunity, and Prof. Dr. Simone Moraes, for her patience, commitment, for always being willing to help me and for her magnificent guidance.

To the coordinators of POSPETRO, Prof. Dr. Antônio Fernando and Prof. Dr. Olivia Maria, for their dedication in carrying out this project and to the teachers of the course, who passed on their knowledge with responsibility and dedication.

To CAPES, for the master's scholarship.

To Queiroz Galvao Exploraçao e Produçao, through the project "Geoenvironmental Diagnosis of Mangrove Zones and Development of Technological Processes Applicable to the Remediation of These Zones: Subsidies for an Impact Prevention Program in Areas with Potential for Petroleum Activities in the Southern Coastal Region of the State of Bahia (PETROTECMANGUE-BASUL)", for funding this research.

To everyone who helped in any way to make this work a reality.

"Each person who passes through our lives passes through alone, because each person is unique and none can replace the other! Each person who passes through our lives passes through alone and doesn't leave us alone because they leave a little bit of themselves and take a little bit of us with them. This is life's most beautiful responsibility and proof that people don't meet by chance."

Charles Chaplin

SUMMARY

Foraminifera are widely recognized as effective tools for identifying ancient and recent environmental variations caused by natural and abiotic factors. Among their many scientific applications, we can highlight their use as bioindicators of environmental stress and in the characterization and biomonitoring of coastal regions. The aim of this study is to characterize the mangrove areas and estuarine channels of the Una, Pardo and Jequitinhonha rivers, on the southern coast of the state of Bahia, using foraminiferal paralogs and geochemical data from the sediment. To this end, two seasonal sampling campaigns were carried out in November 2011 and April 2012, each time collecting 10 samples of bottom surface sediment in the channels and 6 samples of sediment near the *Avicennia* specimens in the mangroves of the estuaries mentioned above. In the first campaign, 770 foraminiferal specimens were obtained from the canals and 763 from the mangroves, of which 12.21% and 4.06% were alive at the time of collection, respectively. In this same campaign, 19 species were recorded (13 found in the canal and 17 in the mangroves), with *Ammonia beccarii, Haplophragmoides wilberti, Miliammina fusca, Trochammina inflatea* and *Trochammina squamata* being common to the three environments studied. In the second sampling period, 125 worms were recorded in the canals and 272 in the mangroves, of which 9.6% and 9.55%, respectively, were alive at the time of collection. During this period, 20 species were identified (14 found in the canals and 17 in the mangroves), of which *A. beccarii, A. tepida, H. wilberti, T. inflata, T. squamata* and *Quinqueloculina fusca* were found in all three study areas. It should be noted that the presence of the planktonic species *Globigerinoides sp.* and *Globigerina pachyderma* was recorded in the Pardo River, which are typical of marine environments far from the coast but which, due to the influence of marine currents and tides, have been transported into the estuary. In addition, the presence of the species *A. beccarii, A. tepida, H. wilberti* and *E. Excavatum* are due to the fact that they are the most tolerant of environmental factors, such as the variations in salinity that occur in the Pardo and Jequitinhonha rivers, as well as being species that prefer environments with high concentrations of nutrients, characteristic of the regions under study, which have a predominance of silt and clay fractions, a textural composition that favors the contribution of organic matter, both in the channels and in the mangrove zones of the three regions studied.

Keywords: foraminifera, Una, Canavieiras, Belmonte, mangrove.

1 INTRODUCTION

The term estuary is used to designate environments that represent a boundary between marine, terrestrial and fluvial environments, communicating the ocean area with coastal zones. They are influenced by natural agents (e.g. climatic, geological, biological, chemical, oceanographic and hydrological events) that occur in the drainage basins and also close to the oceans, acting on the vital processes of organisms and thus interfering with their distribution and abundance in these environments (D'AQUINO et al., 2010; LESTER et al., 2011). In this way, estuarine environments, because they have different physiographic aspects and hydrodynamic circulation patterns, due to the association of the geomorphology of each region with the tidal regime and river discharges, act as a water filter or nutrient exporter for adjacent coastal zones (SCHETTINI et al., 2000).

Among estuarine environments, the mangrove stands out because it is an extremely flooded environment that receives a large amount of sediment, which can come from a variety of sources (such as the decomposition of rocks of various types, or the remains of plants and animals) and is also subject to various types of transportation, ranging from the action of tides and winds to the flow of rivers (COSTA et al., 2011). In addition, the mixing of marine and river waters, as well as the frequent flooding caused by tidal movements, cause major changes in its pH due to the availability of sulphides which favor the precipitation of metals in the reducing conditions of the sediment's surface layer (PEREIRA et al., 2011).

Thus, biogeochemical aspects such as primary production and flocculation of organic matter increase the mangrove's capacity to produce organic matter, making it a place of food and nutrition for various animals and plants (OLIVEIRA et al., 2011). Another important aspect is the fact that its geomorphology helps to fix unstable soils through typical vegetation, since these plants accompany sedimentation, stabilizing coastal and estuarine margins (CUNHA-LIGNON et al., 2009). Therefore, mangroves are among the most important coastal ecosystems, as they are home to a great diversity of birds, insects, fish, molluscs and crustaceans, many of which reside in this ecosystem for their entire life cycle and have great economic and social value, such as oysters, shrimp, sururus, mullets and crabs (KATHIRESAN; QASIM, 2005; OLIVEIRA et al., 2013).

On the other hand, as a result of the development and installation of human activities close to the coastal zones, estuarine channels and mangroves are among the regions most affected by intense demographic and economic pressures, such as deforestation, domestic, agricultural or industrial sewage, landfills, paving, drainage and road construction; and also due to changes in the conditions of estuarine circulation with the construction of dams and reservoirs and channel straightening and dredging for the implementation of port activities, generating ecological changes in these environments (SODRÉ et al., 2002; QUEIROZ et al., 2008; RODRIGUES, 2011; STATHAM,

2012).

Given that these regions are areas of great biological and nutrient productivity, programs for diagnosing, monitoring and managing estuaries are becoming increasingly necessary for the rational exploitation of their natural resources and better environmental monitoring (QUEIROZ et al., 2000; BURFORD, 2012). Thus, the applicability and use of benthic bioindicators, since they are the most exposed to pollution (DONNICI et a., 2012), has intensified, with foraminifera standing out due to their high sensitivity to pollutants, especially non-essential chemical elements (SAMIR, 2000; DEBENAY, 2001; UEHARA-PRADO et al., 2007).

The aim of this study was to **use the characteristics of paralic foraminifera and sediment geochemical data to describe the estuarine channels and mangrove areas of the Una, Pardo and Jequitinhonha rivers, on the southern coast of Bahia,** thus contributing to the formation of a scientific database for the ongoing environmental monitoring of this region.

1.1 FORAMINIFEROS

Foraminifera are protists with a calcareous, proteinaceous, siliceous or agglutinating forehead that have a wide distribution in estuarine and marine environments, occurring abundantly at the most varied latitudes (MURRAY, 1991; KOUKOUSIOURA, 2011). They can be benthic (found on the surface or within the bottom sediment or associated with fixed substrate) or planktonic (inhabiting tropical, subtropical or polar marine waters, with normal salinity, as they are stenohaline) (ZILI et al., 2008; MORIGI, 2009).

Due to the complexity and diversity of their habitats, foraminifera have a wide variety of foods, feeding on small organisms such as bacteria, diatoms and invertebrate larvae that are captured through their pseudopods (DIZ et al., 2006).

The life cycle of these protozoa can vary from one month to one year, in which reproduction can occur both sexually and asexually, resulting in genetic reorganizations that enable them to respond quickly to environmental changes and make them the first organisms in the paralytic community to respond to the effects of environmental stress, whether these are of short or long duration (ARMSTRONG; BRASIER, 2005; PRAZERES, 2007 SARASWAT et al., 2011).

Among the environmental factors that increase the maturation and reproduction of foraminifera, controlling their diversity and distribution, are the intensity and availability of light, temperature, salinity, pH, oxygen content, the chemical composition of the primary and secondary elements in the water, the quantity and type of prey and the turbulence of the water (BOLTOVSKOY et al., 1980; LEVIN et al., 2001; REBOTIM, 2009). Thus, acidic, brackish and poorly oxygenated interstitial waters with sedimentation rich in organic matter, which constitute the muds of mangroves and the clay banks of estuarine margins, offer high stress to the microbiota (DEBENAY, 1990). As a result, the foraminiferal assemblage is very particular and is generally made up of agglutinating *taxa* and/or opportunistic species (BRONNIMANN et al., 1992; DEBENAY et al., 2002). This results in low diversity and evenness indices and high dominance (SEMENSATTO JR.; DIAS-BRITO, 2004).

In addition, the presence of foraminiferal testes with malformations has been reported in areas contaminated by non-essential chemical elements, domestic sewage or various other chemical pollutants, including liquid hydrocarbons (YANKO et al., 1994). Among the ecological factors that influence the deformation of the foreheads of these protozoa, the following stand out: salinity, one of the strongest determinants of the distribution of these microorganisms; pH, as it can prevent the growth of these organisms; and dissolved oxygen content, which determines the distribution of foraminifera in shallow bottom sediments and between sediment grains, especially in situations of shallow water contamination (BARBOSA et al., 2005).

According to Boltovskoy and Wright (1976), since malformations in foraminifera can be caused by natural ontogenic processes or by deleterious effects arising from ecological events (e.g. sudden variations in salinity), only percentages exceeding 1% of deformed individuals are considered significant for determining pollutants (GESLIN et al., 2002).

1.1.2 Foraminifera taphonomy

Taphonomy is the study of preservation processes and how they affect the information in the fossil record. It comprises two broad subdivisions: biostratinomy, which encompasses the sedimentary history of skeletal remains until burial, including the causes of death of a given organism, its decomposition, transportation and burial; and fossil diagenesis, which deals with the physical and chemical processes that alter skeletal remains after burial (SIMÕES; HOLZ, 2000).

In the case of foraminifera, these studies include their state of preservation and color variation, so that from their spatial distribution and the morphometric, morphological and taphonomic analyses of their foreheads, hydrodynamic patterns can be established (LANÇONE et al., 2005) and sediment deposition, erosion and reworking rates can also be interpreted (LEÂO; MACHADO, 1989). Because of this, these organisms are also used in geological studies, since, when they die, their foreheads join the sediment and start to behave like sedimentary grains (KOHO, 2008), so that many studies have been carried out using the patterns of variation in the coloration of foraminiferal foreheads in recent marine sediments (MORAES; MACHADO, 2003; MORAES, 2006; LEMOS JÚNIOR, 2011; ARAUJO et al, 2011; MACHADO et al., 2012).

These protists generally have white or colorless foreheads, but depending on the presence of certain impurities, such as iron, their colors can be altered (MORAES; MACHADO, 2003). Therefore, coloration is a result of the depositional history of the sediment and the structure of the grain, and can indicate:

-	predominance of black grains - a high rate of reworking, with the removal of these black grains from the lower layer to the surface (LEÂO; MACHADO, 1989) or erosion of the oxidizing layer (MORAES, 2006);

-	predominance of brown grains - a slow rate of sedimentation caused by intense, but not rapid, erosion and/or bioturbation, so that the black grains are exposed to the oxidizing action of the surface (MAIKLEM, 1967);

-	predominance of yellow grains - a phenomenon similar to what happens with brown grains, but at an even slower rate (ALMASI, 1978);

-	predominance of white grains - means a very fast rate of deposition, with a lot of new material added to the sediment, or a very slow rate, with extremely exposed and totally oxidized material and/or absence of iron (MACHADO, 1997).

Mottling, on the *other hand,* is related to the structure of the grain, since the coloring agent is preferentially located along the sutures, inside the micropores of the wall of the openings and empty chambers of the testa (LEÂO; MACHADO, 1989). For this reason, one of the few authors to discuss the environmental interpretation of this pattern is Duleba (1994), who suggests that it is a transitional stage between monosulphide and iron hydroxide/oxide; and Moraes and Machado (2003), who interpreted the mottled foreheads (white and yellow) as indicating a transition between two different environmental conditions, such as environmental hydrodynamics and, possibly, exposure of the grains to oxidation.

Thus, variations in the coloration of foraminiferal testes can be used in paleoenvironmental reconstructions (DULEBA, 1994) and to interpret the rate of deposition, erosion and reworking of the sediment (DEBENAY, 1996), since these variations reflect the depositional and diagenetic history of the sediment (TOLER; HALLOCK 1998).

In addition, according to Cottey and Hallock (1988), various patterns of wear can modify the structure and appearance of the foreheads of these organisms after death, so that three types of wear are known: dissolution, abrasion and breakage. Dissolution generally occurs in low-energy environments with sediment rich in organic matter, so that the initially smooth surface of foraminiferal foreheads becomes opaque, rough, vitreous or pulpy until it is perforated; the pores, when present, are enlarged; the surface layer is destroyed and the organic matrix is exposed, making the forehead susceptible to breakage, i.e. completely dissolved (MURRAY; WRIGHT, 1970). This can happen through processes: biochemical, through the action of enzymes in the digestive tracts of polychaetes, crustaceans, gastropods, echinoderms and fish or by exposure to environments rich in algae and bacteria (HICKMAN; LIPPS, 1983); or geochemical, when the microenvironment around the foreheads is intermittently undersaturated (COTTEY; HALLOCK, 1988), being less marked along the sutures and more pronounced on the topographically high parts of the foreheads and in individual chambers, so that a slight abrasion can promote the dissolution of these parts (MURRAY; WRIGHT, 1970).

The other pattern is abrasion, which is very common in high-energy environments and is related to the exposure of the foreheads to the movement of water, being more severe in shallow and relatively exposed environments (COTTEY; HALLOCK, 1988).

Finally, breakage occurs more frequently in high-energy environments (WETMORE, 1987). However, various studies have shown that the dissolution of the outer layer of the foreheads (MURRAY; WRIGHT, 1970); bioerosion - ingestion of foraminifera by marine invertebrates and vertebrates - (SWINCHATT, 1965; HICKMAN;

LIPPS, 1983); and the asexual reproduction of these organisms (ROSS, 1972) can, together or not, weaken the forehead by causing its periphery to break or one or more of its chambers to be lost as efficiently as the movement of water (COTTEY; HALLOCK, 1988).

Thus, break-up does not indicate any specific depositional environment because it can result from various physical, chemical or biological processes, but in low-energy environments dissolution prevails, while abrasion occurs in shallow, open-platform environments (COTTEY; HALLOCK, 1988).

Therefore, by studying the species composition and taphonomic analysis of foraminifera, it is possible to infer the pattern of estuarine circulation, identify bodies of water and understand local sedimentary dynamics (ARMYNOT DU CHATELET et al., 2008). In addition, it helps to understand environmental problems, both natural and man-made, by allowing these events to be recorded in the geochemical and/or isotopic composition of their samples (FRONTALINI; COCCIONI, 2008).

1.2 Characterization of the areas studied

The study area corresponds to the channels and mangrove areas in the estuaries of the Una, Pardo and Jequitinhonha rivers, on whose banks lie the municipalities of Una, Canavieiras and Belmonte, respectively, located on the Dendê Coast, on the southern coast of the state of Bahia (Figure 1.1).

This region is characterized by a hot and humid climate, with well-distributed rainfall, with a predominance of rainy months (October to April), with only 1 to 2 drier months (May and August)

(DE PAULA et al., 2012). It has a relative humidity of over 85%, which decreases inland (BARRETO, 2005).

From an environmental point of view, the region is characterized by the presence of a series of coastal ecosystems, such as areas of sandbanks, sandy beaches, dunes, rocky coasts, oceanic islands and marshes, as well as remnants of Atlantic forest and the presence of estuaries, where the mangrove ecosystem is found, in which the species *Rhizophora mangle*, *Rhizophora racemosa* and *Avicenia* spp grow (PERH- BA, 2003).

Figure 1.1 - Location map of the municipalities of Una, Canavieiras and Belmonte and geomorphological units of the study area

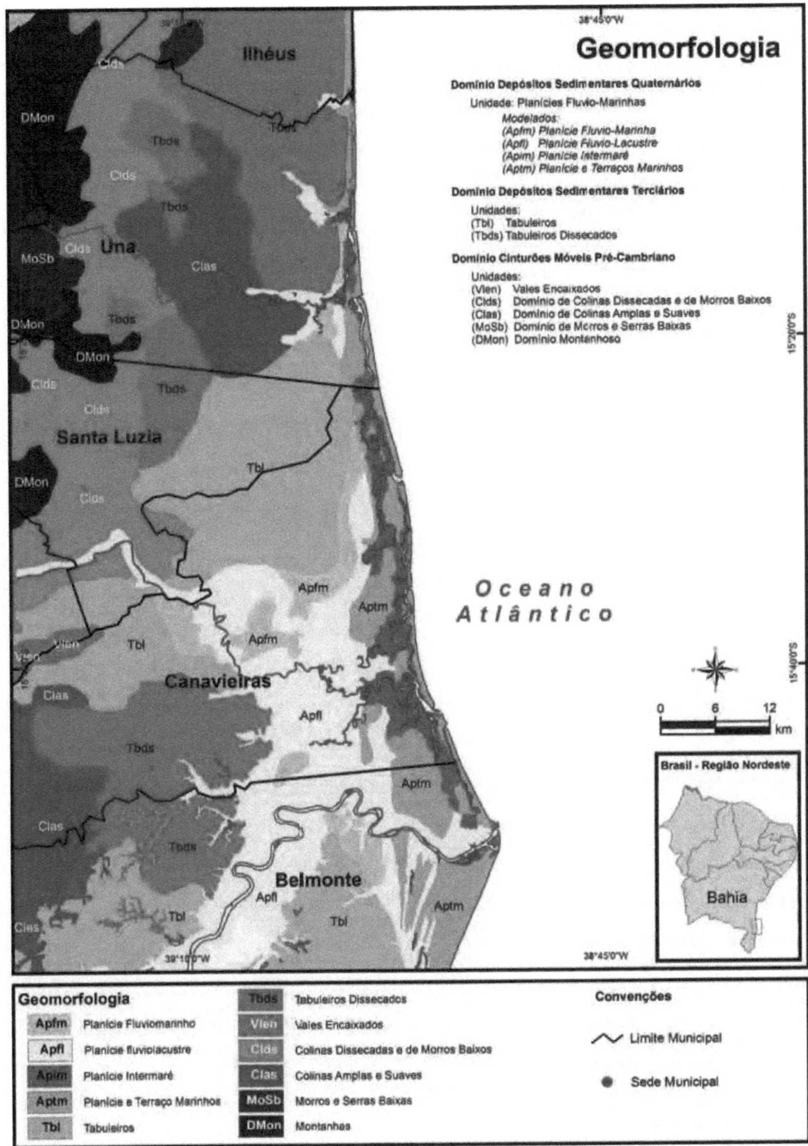

Prepared by: Adriano de Oliveira Vasconcelos

Source: VASCONCELOS; CELINO, 2014, p. 32.

The mangrove areas are cut by tidal channels and form an estuarine complex that connects the mouths of the Pardo and Jequitinhonha rivers. In addition to the tidal channels, the strong erosive action of the Pardo and Jequitinhonha rivers promotes the expansion of the river terraces, which

11

have been occupied by the forest (SANTOS, 2007).

The Una River rises in the Serra de Sao Roque within the boundaries of the municipality of Arataca at a height of 620 m above sea level, and is part of the Atlantic Forest, covering 94 km from its source to its mouth (HIGESA ENGENHARIA, 1996a), where there is a population of 22,989 inhabitants, distributed over an area of 1,177.440 km^2. Its remaining mangroves occupy an area of 0.01853 km^2, representing 1.6% of the municipality's total area (IBGE, 2013). In the area covered by the Una river basin there are two small hydroelectric plants, Juçari and Dendhevea (ROCHA, 1976), whose energy supply is used for activities such as flour mills, sarraria, rubber manufacture, domestic use, among others (PDTS, 2010). Other important activities in the region are the cocoa plantations that began in the 18th century (only fully developed in the 19th century), together with extractive activities, sugar cane cultivation, subsistence farming and shrimp farming (MANAGEMENT PLAN, 1997).

The Pardo river basin rises in the municipality of Rio Pardo de Minas, at an altitude of 750 m, in the state of Minas Gerais, and is 565 km long (XAVIER, 2009). It drains 27 municipalities in the southeast of the state of Bahia (PERH-BA, 2003), ending its course in the municipality of Canavieiras (15°41'S and 38°57'W), which has 33,570 inhabitants in an area of 1,326.931km^2, whose mangroves cover 0.07403 km^2 (5.4% of the municipality's total area) (IBGE, 2013). The municipality of Canavieiras (BA) has an important economic activity in artisanal fishing and has significant natural wealth, such as its extensive coastline and estuaries, extensive mangrove areas and a great diversity of species of fauna and flora (AGUIAR et al., 2011). In this sense, the Extractive Reserve (RESEX) was created in June 2006, in coastal areas and also in areas where some resorts are expected to be installed, in order to restore and/or conserve the already degraded environment and that threatened with degradation (AGUIAR et al., 2011). Thus, a number of environmental impacts on the mangrove forest in the municipality of Canavieiras can be highlighted: deforestation of the mangrove forest within the Resex; implantation of shrimp farms on Apicum areas; silting up of the mangrove forest; drainage of various wetlands (marshes and lagoons); human pressures resulting from disorderly occupation; absence of a sanitary sewage system (AGUIAR et al., 2012). In addition, there are also problems caused by agricultural and livestock activities carried out in the transition areas between mangroves and sandbanks.

However, it should be noted that the main causes of environmental stress come from shrimp farming around the mangrove (DIAS et al., 2012).

With a length of 920 km, the Jequitinhonha River rises in the Serra do Espinhaço, south of the city of Diamantina, in Minas Gerais, at an altitude of around 1,200 m, and flows into the Atlantic Ocean, in the municipality of Belmonte, on the southern coast of Bahia, which has a population of 23,471 inhabitants, distributed over an area of 1,970.142 km^2 (IBGE, 2013).

In the study area, the impacts are caused by pastoral and agricultural activities, irregular occupation, the dumping of domestic waste, and intense silting that compromises navigation, especially on the Jequitinhonha River. The region has also been pressured by activities linked to shrimp farming, oil exploration and forestry, which have probably directly affected coastal environments (FARIAS, 2007). In addition, poor land use is one of the main factors responsible for the degradation of water quality, and this interference is evidenced by changes in the values of total manganese, dissolved iron, turbidity, total suspended solids and total aluminum (IMGA, 2010).

2 MATERIALS AND METHODS

In order to meet the objectives of this work, two collections were carried out: one during the rainy season, from 25 to 27/11/2011; and the second during the dry season, from 21 to 23/04/2012.

From these collections, a sample set of 96 samples was obtained (Figure 2) as follows:

• 30 samples of bottom surface sediment in the estuarine channel of the Una, Pardo and Jequitinhonha rivers (10 samples in each channel per campaign, totaling 60 samples) obtained using a van Veen; and

• 18 samples of surface sediment collected manually with the aid of a polyethylene spoon near specimens of *Avicennia* in mangrove areas in the municipalities of Una, Belmonte and Canavieiras (6 samples in each mangrove per campaign, totaling 36 samples).

In both cases, the samples were placed in plastic vials with screw caps and Rose Bengal dye was added to fix the individuals captured alive at the time of collection.

Still in the field, the geographical coordinates of the points were recorded using a GPS and the physico-chemical parameters of the water were measured using a multiparameter probe.

Figure 2.1 - Manual sampling of mangrove sediment in the Una river estuary, during the first campaign (nov/2011)

2.1 LABORATORY PROCEDURES

In the laboratory, geochemical analyses were carried out on the sediments, i.e. granulometric, metal and nutrient analyses. With regard to foraminifera, taphonomic signs were recorded on the shells of the species.

2.1.2 Foraminifera analysis

In the Laboratories of the Nùcleo de Estudos Ambientais - NEA (Institute of Geosciences/Federal University of Bahia), all the samples were washed under running water and then taken to the oven to dry at a temperature of 60°C for three days.

Afterwards, about 3g of dried sediment was poured into beakers and trichloroethylene was added to separate the organic part (supernatant) from the inorganic part of the sediment (Figure 3). All the floated material was then placed on filter paper and left in the oven for approximately 5 minutes to dry.

After removal from the oven, each piece of filter paper containing the organic matter, often including abundant plant remains, was stored in plastic bags until it was observed under a stereomicroscope (Figure 4), when the samples were sorted and identified in the Foraminifera Studies Group Laboratory (LGEF/UFBA).

The foraminifera were identified using specialized literature, based on Loeblich and Tappan (1988). During this procedure, information on their coloration and state of preservation was recorded in accordance with Moraes and Machado (2003). The occurrence of forehead anomalies was also recorded, when present.

2.1.3 Sediment analysis of the canals and mangrove area

The sediment collected was analyzed by researchers from the project "Geoenvironmental Diagnosis of Mangrove Zones and Development of Technological Processes Applicable to the Remediation of These Zones: Subsidies for an Impact Prevention Program in Areas with Potential for Petroleum Activities in the Southern Coastal Region of the State of Bahia (PETROTECMANGUE-BASUL)". Thus, the data for the analysis of granulometry, assimilable phosphorus, organic matter, total nitrogen and metals in the sediment of the channels of the Una, Pardo and Jequitinhonha river estuaries were obtained from Escobar (2013; Celino et al., 2014; Escobar et al., 2014) and those for the mangrove zones can be found in Cruz (2012), with the methodological procedures summarized in the following table.

Figura 2.2   - Procedure for flotation of foraminiferal samples from the estuaries and channels of the Una, Pardo and Jequitinhonha rivers

Figura 2.3 - Procedure for identifying foraminifera, using a stereomicroscope, in the estuaries of the Una, Pardo and Jequitinhonha rivers

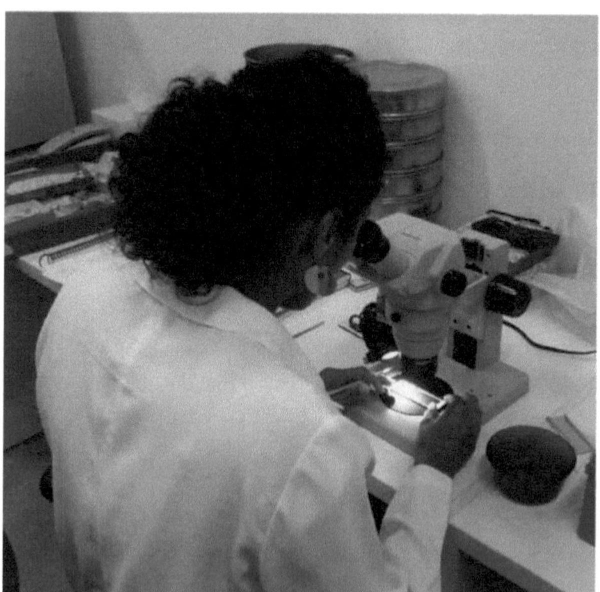

Table 2.1 - Analysis methods for the 32 surface and bottom sediment samples collected in the mangrove zone and in the Una, Pardo and Jequitinhonha channels.

Parameters	Analysis/Determination
Granulometry	Sample pre-treatment according to Embrapa (1997). Analysis carried out using a particle analyzer with laser diffraction (mod. Silas 1064) and data processing using

	GRADSTAT software.
Organic Matter - M.O.	Organic matter - O.M. Determination of total organic carbon using the Walkey - Black method (1947). To calculate O.M., the value C.O.x 1.724 was taken (EMBRAPA, 1997).
Total Nitrogen - NT	Total Nitrogen - NT. The Kjeldahl method was used, as recommended by Embrapa (1997).
Assimilable Phosphorus - P	Assimilable phosphorus - P. It was determined using the methods of Grasshoff et al. (1983) and Aspilla (1976).
Partial Metal Extraction	Partial extraction of metals. 65% HNO3 was used, and it was carried out according to methodology D 5258-92 (ASTM, 1992).
Total Metal Extraction	Total extraction of metals. HNO3 a, HCl 37%, HF 40% were used and the extraction was carried out according to EPA methodology 3052 (USEPA, 1996).

Source: adapted from SANTOS, 2013

2.1.4 Statistical analysis

For the descriptive analysis of the foraminiferal fauna, relative abundance and relative frequency of occurrence were calculated (AB'SABER et al., 1997). The relative frequency (F) is defined as the ratio between the number of individuals in a category (n) and the total number of individuals in all categories (T), expressed as a percentage, namely

$$F = n.\frac{100}{T}$$

According to Dajoz (1983), the following classes were adopted: main (abundances >5%), accessory (4.9-1%) and trace (<1).

The frequency of occurrence is based on the number of occurrences of a category (p) in relation to the total number of samples (P) (AB'SABER et al., 1997).

$$C = p.\frac{100}{P}$$

Dajoz's (1983) classification was used to assess the frequency of occurrence, i.e. constant (occurrences >50%), accessory (49-25%) and accidental (<24%).

In addition, with the help of the Primer 6.0 program, the indices of richness (Margalef index), evenness (Pielou index) and diversity (Shannon-Wiener index) were calculated.

(CLARKE; WARWICK, 2001). In order to establish the relationship between the distribution of foraminifera and environmental variables, the faunal results were combined with those of the physico-chemical parameters and the granulometric and geochemical analyses to create a data matrix, from which principal component analyses were carried out using the Statistica 7.0 program (StatSoft, 2007).

3 BIOGEOCHEMICAL CHARACTERIZATION OF THE CHANNEL AND THE MANGROVE ZONE OF THE INUANA RIVER ESTUARY, SOUTH COAST OF THE STATE OF BAHIA, BASED ON PARALLIC FORAMINIFERS AND GEOCHEMICAL SEDIMENT DATA

Summary

Species associations and foraminiferal test characteristics were related to the levels of trace metals in the sediment of the mangrove zone and the estuarine channel of the Una River, south coast of Bahia, in order to assess whether the levels of these elements are affecting the microfauna. Two seasonal sampling campaigns were carried out (November/2011 and April/2012) to collect, in each campaign, 10 samples of surface sediment at the bottom of the estuarine channel and 6 samples of sediment near specimens of *Avicennia* in the mangrove, totaling 32 samples. In the estuarine channel, 21 foraminiferal tests were obtained in the first campaign (28.58% of the specimens were collected alive; and none of the tests were malformed) belonging to 5 species, of which *Haplophragmoides wilberti*, *Ammonia beccarii* and *Elphidium excavatum* stand out. In the second sampling, 96 testes (11.45% live; 0% anomalies) of 11 species were recorded, with *T. inflata, H. wilberti, Ammonia beccarii* and *Elphidium excavatum* predominating. In the mangrove zone, in the first campaign, 483 testes were obtained (2.08% live; 0.0% anomalous) from 10 species, the most frequent being *A. beccarii, H. wilberti, T inflata* and *T. squamata*. In the second campaign, only 86 tests were recorded (10.46% live; 0.0% anomalies), with *H. wilberti, A. tepida, T. inflata* and *E. excavatum standing* out. In the channel, the concentrations of all the elements are below the threshold for adverse effects on biota, but in the mangrove the levels of lead and cadmium exceeded the reference limits. In the estuarine channel, the negative correlation of live individuals with MO>OD during the rainy season indicates their presence in places with high consumption of dissolved oxygen and organic matter. In the dry season, organisms were related to OD>MD>FS levels due to increased precipitation of fine sediment and nutrient concentrations. On the other hand, the number of dead individuals showed a negative correlation with the coarse sand content in both campaigns due to the influence of hydrodynamic energy. In the mangrove, on the other hand, the distribution of live foraminifera did not correlate with any of the parameters analyzed, suggesting that it was influenced by salinity levels; while that of dead individuals was related to pH>VS due to their preferential deposition in conditions of low hydrodynamic energy. In the dry period, living and dead foraminifera were related to pH due to the increase in salinity values.

Keywords: Foraminifera, Una river, mangrove, estuary

3.1 INTRODUCTION

Estuaries are coastal regions where there is a measurable dilution of marine waters by the dulcicoles, thus communicating the continental drainage to the oceanic area (LEGORBURU et al., 2013). The geomorphology of these environments, together with the tidal regimes and river discharges, provide distinct patterns of water circulation in each estuary, allowing these environments to function as filters and exporters of nutrients to the adjacent coastal regions (BRICKER et al., 2008). On the other hand, they are intensely subject to environmental impacts resulting from anthropogenic activities, such as urban, industrial, chemical, agricultural, fishing and recreational activities, which affect the natural functioning of this ecosystem (DONNICI et al., 2012).

Another environment sensitive to human actions is the mangrove zone which, from an environmental point of view, is characterized as a tropical ecosystem that contains plant communities typical of flooded environments with great variations in salinity, as well as acting as a nursery and nutrient source for a huge number of animal and plant species (SATYANARAYANA et al., 2013).

Given that estuarine and mangrove regions are home to a large number of species and organisms, many of which have a high economic value associated with fishing and leisure resources (MAIA et al., 2012), studies of environmental impacts on these sites have become commonplace around the world (e.g. TEODORO et al., 2010; SANTOS et al., 2012; 2014). Thus, there are several ways to study and characterize the conditions of these ecosystems, but the most effectively used is the one that combines granulometric and geochemical analyses of the sediment with studies of bioindicators (DAUVIN, 2007). Among these, paralic foraminifera stand out for their high sensitivity to physical and chemical changes in the environment, so studies with these organisms have been increasingly carried out in all regions of the world in order to understand the natural dynamics of ecosystems and also detect anthropogenic changes promoted in these environments (FRONTALINI et al., 2008; TEODORO et al., 2009; 2010; 2011; EICHLER et al., 2012).

In view of the above, the aim of this study was to use the characteristics of paralic foraminifera and sediment geochemical data to assess whether trace metal levels are affecting the microfauna of the estuarine channel and mangrove zone of the Una River, on the southern coast of Bahia.

3.2 MATERIALS AND METHODS

The study area corresponds to the channel and mangrove area of the Una River estuary on the southern coast of the state of Bahia (Figure 3.1). The Una River rises in the Serra de Sao Roque within the boundaries of the municipality of Arataca at 620 m above sea level and runs for 94 km to its mouth (DE PAULA et al., 2012), where the municipality of Una is located with a population of 22,989 inhabitants, distributed over an area of 1,177.440 km^2 (IBGE, 2014).

From an environmental point of view, the region is characterized by the presence of a series of heavily impacted coastal ecosystems, such as areas of sandbanks, sandy beaches, dunes, rocky coasts, oceanic islands and marshes, as well as remnants of Atlantic forest and the presence of estuaries, where the mangrove ecosystem is found, where the species *Rhizophora mangle*, *Rhizophora racemosa* and *Avicenia* sp grow (BAHIA, 2010).

This region can experience rainfall in excess of 2000 mm, so the predominance of rainfall characterizes the climate as tropical rainy (with a brief dry period), increasing the water yield of the Una River basin (RECIFE, 2011). The wet season comprises the months between April and August (in general, the months of June and July have the highest flows, with average annual flows of 16m^3 sec^{-17}), while the dry season is from October to March (DE PAULA et al., 2012). However, during the sampling months, there was an anomaly in the region's rainfall regime, with the highest index (332 mm per month) recorded during the first campaign, behaving like a rainy period, while in the second sampling the rainfall rate was reduced (50 mm per month), characterizing it as a dry season (ESCOBAR et al., 2014), which certainly influenced the values of the parameters recorded during the collections. Therefore, for the purposes of this work only, the first campaign will be considered to have been carried out during the "rainy season" and the second during the "dry season".

3.2.1 Sampling procedure: sample collection and analysis

Two sampling campaigns were carried out - one during the rainy season (25/11/2011) and the other during the dry season (23/04/2012) - in which 10 samples of surface sediment from the bottom of the Una River estuarine channel were obtained using a van Veen. In addition, in the mangroves, six samples of the sediment near specimens of *Avicennia* were collected using a polyethylene spoon.

In both cases, two groups of samples were collected: one for the study of foraminifera and the other for granulometric and geochemical analysis of the sediment. The samples from the first group were packed in plastic bottles with screw caps and Rose Bengal dye was added to fix the individuals captured alive at the time of collection, and they were kept refrigerated until the time of analysis. The samples from the second group, were stored in aluminum containers and frozen until processing in the laboratory.

Figure 3.1 - Procedure Map showing the location of the Municipality of Una and the sampling points in the canal and mangrove swamp. Extracted from Queiroz and Oliveira (2013)

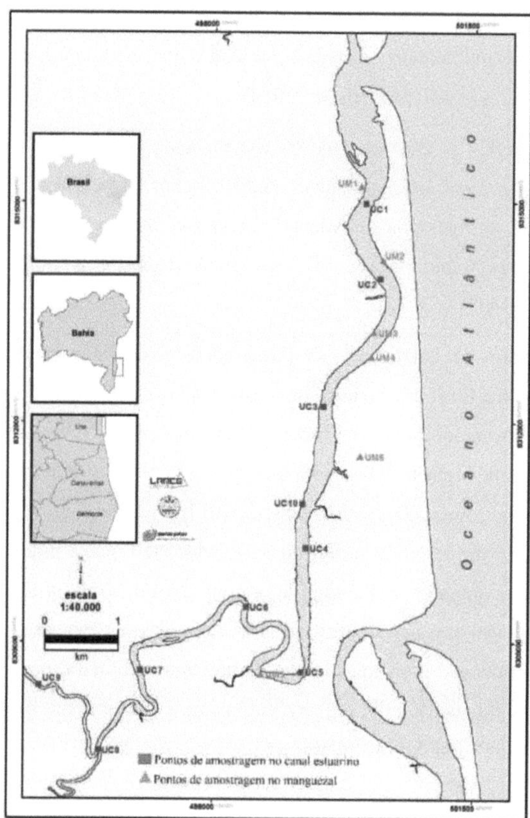

Source: QUEIROZ and OLIVEIRA, 2013

During the collections, the physico-chemical parameters, temperature, dissolved oxygen (DO), of the estuarine channel water and mangrove interstitial water and hydrogen potential (pH), and oxidation potential (Eh) of the channel and mangrove sediments were recorded using multiparameter probes.

3.2.2 Foraminifera fauna analysis

In the laboratories of the NEA (Institute of Geosciences/Federal University of Bahia), the samples intended for the study of foraminifera were washed under running water and taken to the oven at 60°C. After drying, about 3g of the sediment was poured into beakers, into which trichloroethylene was added to separate the sediment samples by flotation. The supernatant was then placed on filter paper and left in the oven for approximately 5 minutes to dry.

In the Laboratory of the Foraminifera Study Group - LGEF, using a stereomicroscope, the specimens were removed from the filter paper and fixed with organic glue on microfossil slides.

The foraminifera were identified using specialized literature, based on Loeblich and Tappan (1988). During this procedure, information was recorded on their coloration and state of preservation according to Moraes and Machado (2003), as well as the occurrence of anomalies in the tests.

3.2.3 Sediment analysis

The results of the analysis of the sediment from the Una River channel and mangrove swamp were provided by NEA researchers, so the procedures and data from the analysis of granulometry, assimilable phosphorus, organic matter, total nitrogen and metals can be found in Celino et al. (2014), Escobar et al. (2014) and Cruz (2012).

The criteria for trace metal levels in sediment established by Brazilian environmental legislation can be found in Resolution No. 454 of the National Environment Council (Brazil, 2012), but for comparison purposes, the parameters of the *Canadian Environmental Quality Guidelines (*CEQG), *Canadian Council of Ministers of the Environment (*CCME, 1998) were also adopted.

3.2.4 Statistical analysis

For the descriptive analysis of the foraminifer fauna, the relative abundance and frequency of occurrence were calculated and the following classes were adopted according to Dajoz (1983): main (abundances >5%), accessory (4.9-1%) and trace (<1); and constant (occurrences >50%), accessory (49-25%) and accidental (<24%). In addition, the richness (Margalef index), evenness (Pielou index) and diversity (Shannon-Wiener index) indices were calculated using the Primer 6.0 program (CLARKE; WARWICK, 2001).

In order to establish the correlation between foraminiferal distribution and environmental variables, the faunal results were combined with data on physical-chemical, granulometric and geochemical parameters to create a data matrix, from which principal component analysis was carried out using the Statistica 7.0 program (StatSoft, 2007).

3.3   RESULTS AND DISCUSSIONS

The results of the physicochemical parameters monitored in this study showed a rainfall anomaly in the months of the sampling, so that the rainfall was higher (332mm per month) during the first campaign (Nov 2011), which was characterized as a rainy season, while in the second sampling (Apr 2012) the rainfall rate was reduced (50mm per month), characterizing it as a dry season (ESCOBAR, 2013), which certainly influenced the values of salinity, temperature and other parameters recorded during the collections. Therefore, for the purposes of this study only, the first campaign will be considered to have been carried out during the "rainy season" and the second during the "dry season".

3.3.1 Una River estuarine channel: physico-chemical parameters

As expected, temperature values were lower in the rainy season (25.1 to 25.7 °C in Nov/2011) than in the dry season (27.2 to 29.0 °C in Apr/2012) (Figure 3.2 and Table 3.1).

In both campaigns, the estuarine waters of the Una River channel were well oxygenated (9.3 to 26.0 mg/L^{-1} in the first campaign and 5.2 to 10.9 mg/L^{-1} in the second campaign) (Figure 3.2), and the lower values recorded in the dry season may be related to the reduction in water volume and the increase in temperature, as well as the consumption of oxygen through the decomposition of organic matter (Table 3.1), respiratory processes of aquatic organisms and nitrification and oxidation of metal ions (CELINO et al... 2014), 2014).

The pH values of the sediment were alkaline to slightly neutral (5.2 to 7.1 in the first campaign and 5.9 to 7.3 in the second) (Figure 3.2). In the first campaign, when analyzing the location of the collection points in the estuary, slightly more acidic conditions were observed from point 7 onwards, confirming the expectation that the points near the mouth are more influenced by the marine system than the points upstream, which are under strong fluvial influence (VEIGA, 2010), but this pattern was not observed in the second campaign, when rainfall was lower.

For the sediment Eh data, in the first campaign, negative values were obtained up to point 6, configuring it as a reducing environment, while points 7 to 10 were under oxidizing conditions. In the second campaign, only points 8 and 9 had negative Eh values (Figure 3.2).

3.3.2 Sediment analysis

In both samples, coarse sand was the predominant granulometric fraction, indicating a condition of greater hydrodynamic energy in the channel, although the percentages of the fine sand fraction were also high (Figure 3.2).

In the dry season, there was an increase in assimilable phosphorus (12.0mg/L to 124.7mg/L in the first campaign and 2.4mg/L to 83.5mg/L in the second) and total nitrogen (0.0mg/L to 2.6mg/L and 0.5mg/L to 10.2mg/L); and a reduction in organic matter (2.1mg/L to 4.7mg/L and 0.1mg/L to 4.5mg/L) (Table 3.1).

Comparing the levels of the elements analyzed with the limits established by CONAMA (National Environmental Council) Resolution No. 454/2012 (BRASIL, 2012) and by the *Canadian Council of Ministers of the Environment* (CCME, 1998), it can be seen that the concentration of Ni, Cr, Cu, Pb and Cd in the sediment is low and, therefore, should not be causing adverse effects on the biota of the Una River channel (Table 3.2). However, these standards do not set limits for Mn and Fe, which makes it impossible to assess the effect of these elements.

With regard to the sources of these metals, Escobar et al. (2014) state that Ni and Pb concentrations are higher in the rainy season (Table 3.2) due to the influence of precipitation, although the variability in the distribution of these elements between the sampling stations indicates the existence of point sources. These same authors recorded higher concentrations of Fe, Mn, Cr, Cu and Cd in the suspended particulate matter (SPM) in the dry season and therefore attributed the increase in the levels of these metals in the sediment to the increased deposition of SPM in this period, thus suggesting agricultural activities and the dumping of untreated sewage and garbage as possible sources of these metals in the Una river estuary.

Figure 3.2 - Values of the physicochemical parameters temperature, dissolved oxygen (D.O.) of the water and hydrogen potential (pH), oxidation potential (Eh) and granulometric fractions of the sediment of the estuarine channel of the Una River, relative to the campaigns of Nov/2011 (rainy season) and Apr/2012 (dry season).

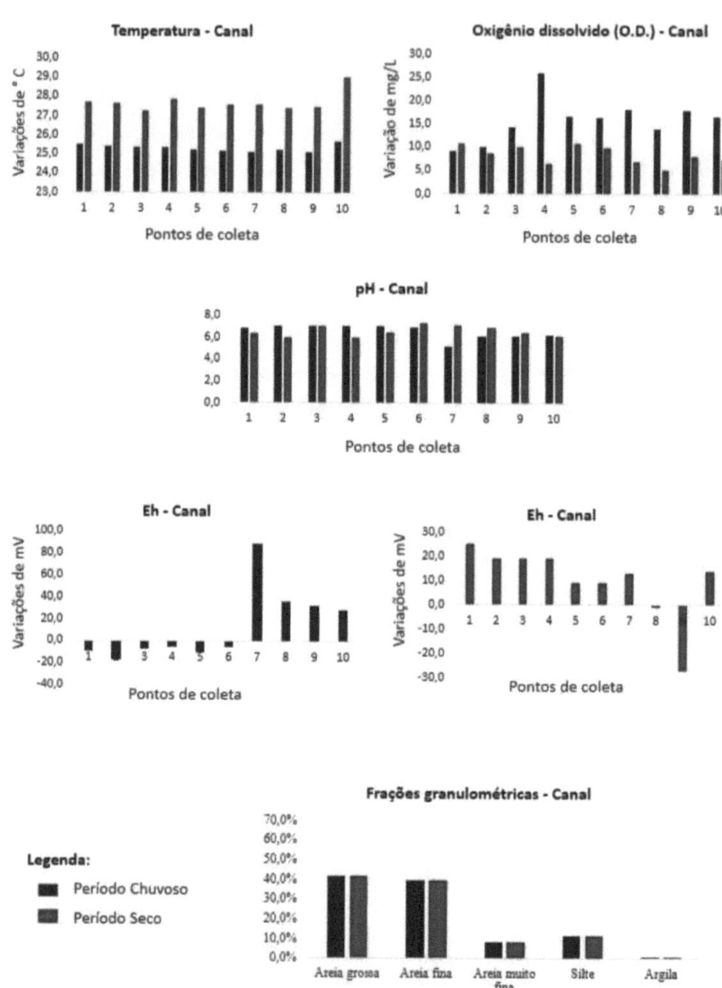

1.1.3 Foraminifera fauna

In the Una river channel, during the rainy season (Nov/2011), 21 testas were obtained, of which only 6 individuals were alive. Almost all the fish were found at points 1 and 2, but no anomalous individuals were recorded. Only 5 species were identified at this station, with *Haploprhagmoides wilberti* (42.9%), *Ammonia beccarii* (38.1%) and *Elphidium excavatum* (9.5%) being the main species (Table 3.3).

During the dry season, 96 individuals were obtained, of which 11 were alive. Individuals were only recorded up to point 4, but all the foreheads were normal. In this campaign, 11 species were identified, of which *Trochammina inflata* (37.5%), *H. wilberti* (24.0%), *A. beccarii* (19.9%) and

26

Elphidium excavatum (5.2%) were considered the main ones (Table 3.3).

The species of the genus *Ammonia* and *Elphidium* are typical of seas and oceans, but can colonize estuarine environments. *Haploprhagmoides* and *Trochammina* are common in estuarine waters with low salinities (MURRAY, 2006). Bearing in mind that the concentrations of metals in the sediment were low, the non-occurrence of foraminifera from point 3 in the rainy season and point 5 in the dry period is possibly due to low salinity values making it impossible for even the most resistant species to survive, which can be corroborated by the increase in the number of individuals, species and occurrence during the dry season (Table 3.3) when it is expected that there will be an increase in marine influence in the estuary. However, no salinity values were measured in the estuarine channel during the sampling.

The richness values ranged from 0.8 to 1.1 in the rainy season and from 1.0 to 2.7 in the dry season and the evenness values were all above 0.5, suggesting the absence of dominance in both samples (CLARKE; WARWICK, 2001). Diversity values ranged from 0.7 to 1.4 (Table 3.3), which are considered low when compared to those obtained by Gomes (2010) in the Jacuipe River estuary, BA (1.0 to 4.4) and by Teodoro (2009 and 2010) in the Sao Sebastiao (0.3 to 2.8) and Araça (2.5 to 3.0) Canals, SP.

3.3.4 Taphonomy of the foreheads

In both campaigns, most of the testes were white/colored (54.55% in the first campaign and 77.08% in the second) or mottled (22.73% and 12.50%). As for wear and tear, the majority were normal (31.82% and 33.33%), but the percentages of dissolution (27.27% and 33.33%) and breakage (27.27% and 23.81%) were significant (Figure 3.3).

The dominance of white and normal foreheads is indicative of a very rapid rate of deposition, with a lot of new material being added to the sediment (MORAES; MACHADO, 2003; MACHADO et al., 2012), especially during the dry season when salinity is expected to rise due to the reduction in river inputs to the estuary.

Table 3.1 - Physico-chemical, granulometric and nutrient data from the Una River estuarine channel for the first (I - Nov/2011) and second (II - Apr/2012) sampling campaigns and limits of CONAMA Resolution No. 357/2005 (Brazil, 2005)

Samples	Water parameters		Sediment Analysis									
	T (°C)	OD (mg/L)$^{-1}$	pH	Eh (mV)	GA (%)	PA (%)	MFA (%)	S (%)	A (%)	NT (mg/L)	P (mg/L)	MO (mg/L)
CUI1	25,5	9,3	6,9	-9	79,7	14,5	3,4	1,8	0	2,6	124,7	2,1
CUI2	25,4	10	7	-17	9,1	81,1	6,1	3,7	0	0	12,4	4,4
CUI3	25,3	14,4	7,1	-7	0,2	0	25,1	73,6	0	0	12	4,3
CUI4	25,3	26	7	-5	36,1	53,9	5,9	4,1	0	0	12,9	4
CUI5	25,2	16,7	7	-10	91,1	7,8	0,4	0,5	0	0	13,7	4,7

Samples	T	O.D.	pH	Eh	AG	AF	AMF	S	A	P	NT	
CUI6	25,2	16,4	6,9	-5	55,6	38,7	3,3	2,4	0	0	13,6	4,6
CUI7	25,1	18,2	5,2	89	40	33,5	14,6	11,8	0	0	14,1	4
CUI8	25,2	14	6,1	36	29,1	62	6,2	2,7	0	0	13,7	4,4
CUI9	25,1	18	6,2	32	54,5	37	5,4	3	0	0	14	4,4
CUI10	25,7	16,7	6,2	28	43,9	45,7	6,6	3,8	0	0	13,6	4,4
Average	25,3	15,96	6,57	13,2	43,94	37,41	7,7	10,73	0	0,28	24,46	41,3
CUII1	27,7	10,7	6,4	25	10	81	4,6	4,4	0	8,6	83,5	4,5
CUII2	27,6	8,6	6	19	2,7	63,5	13,4	20,3	0,1	1,4	17,2	0,4
CUII3	27,2	10	7,1	19	22,3	69,7	5,3	2,7	0	1,9	9,8	0,4
CUII4	27,8	6,5	5,9	19	54,9	8,4	13,2	23,4	0,1	0,5	68,9	0,2
CUII5	27,4	10,9	6,4	9	15,2	77,9	4,6	2,3	0	10,2	67,8	3,8
CUII6	27,6	9,8	7,3	9	14,3	78,4	4,9	2,4	0	1,2	69,6	0,3
CUII7	27,6	6,9	7,1	13	24,6	65,9	6,6	2,9	0	1,1	2,4	0,2
CUII8	27,4	5,2	6,9	-1	23,8	65,5	7	3,7	0	1,6	19,8	0,3
CUII9	27,5	8	6,4	-27	46,5	47,7	3,6	2,2	0	3,5	36,9	0,1
CUII10	29	7,4	6,2	14	66	30,4	1,7	1,9	0	3,9	33,8	0,4
Average	27,7	8,4	6,59	9,9	28,02	58,85	6,49	6,61	0,03	3,39	40,96	1,07
CONAMA												
SWEET	<40,0	5	-	-	-	-	-	-	-	-	-	-
SALINE	-	6	-	-	-	-	-	-	-	-	-	-
SALOBRA	-	5	-	-	-	-	-	-	-	-	-	-

Legend: CUI1 = Una river channel from the first campaign; T = Temperature; O.D. = Dissolved Oxygen; pH = Hydrogen Potential; Eh = Oxirreduction Potential; AG = Coarse Sand; AF = Fine Sand; AMF = Very Fine Sand; S = Silt; A = Clay; P = Phosphorus; NT = Total Nitrogen; CONAMA = National Environment Council.

Table 3.2 - Concentrations of metals (mg. Kg^{-1}) in the sediment of the estuarine channel of the Una River, concerning the first (I - Nov/2011) and second (II - Apr/2012) sampling campaigns and reference values of CONAMA Resolution No. 454/2012 (Brazil, 2012) and the *Canadian Council of Ministers of the Environment* (CCME, 1998).

Samples	Ni	Mn	Fe	Cr	Cu	Pb	Cd
CUI1	10,75	123,15	17597,2	19,67	5,04	8,22	0,02
CUI2	4,17	14,1	1528,83	2,34	0,01	0,85	0,00
CUI3	1,94	10,33	1746,76	0,05	0,01	0,57	0,00
CUI4	0,01	0,00	0,09	0,05	0,01	0,01	0,00
CUI5	4,28	25,42	1157,12	0,29	0,01	0,31	0,03
CUI6	0,01	1,83	561,61	0,05	0,28	0,51	0,05
CUI7	0,01	6,86	1212,83	0,11	0,01	0,47	0,03
CUI8	0,01	10,45	532,49	0,05	0,01	0,25	0,01
CUI9	0,01	11,14	925,75	0,05	0,01	0,46	0,01
CUI10	0,01	48,85	1850,43	2,66	0,01	0,73	0,01
Average	2,12	25,21	2711,31	2,53	0,54	1,24	0,02
CUII1	0,68	21,92	1505,34	2,15	0,41	0,83	0,39
CUII2	3,03	62,03	7241,09	9,57	2,34	2,50	0,56
CUII3	0,18	15,47	1430,48	1,49	0,18	1,14	0,46
CUII4	0,16	6,33	1651,25	0,99	0,08	0,50	0,44
CUII5	1,12	27,95	3440,94	4,02	0,87	1,38	0,49
CUII6	0,01	2,80	1087,72	1,51	0,43	0,80	0,38
CUII7	0,01	9,46	1060,95	0,98	0,06	0,66	0,43
CUII8	0,02	118,52	1641,12	2,00	0,89	0,01	0,49
CUII9	0,38	67,93	4062,48	2,80	1,13	0,01	0,48
CUII10	0,70	197,89	5808,78	4,32	1,80	0,01	0,53
Average	0,63	53,03	2893,02	2,98	0,82	0,78	0,47
LD	0,0041	0,0011	0,0258	0,0163	0,0033	1,49	0,0009
CONAMA N1	20,9	n.a.	n.a.	81	34	46,7	1,2
CONAMA N2	51,6	n.a.	n.a.	370	270	218	7,2
CEQG ISQG	n.a.	n.a.	n.a.	52,3	18,7	30,2	0,7
CEQG PEL	n.a.	n.a.	n.a.	160	108	112	4,21

Legend: LOD = Limit of Detection; N1 and ISQG= threshold below which there is a lower probability of adverse effects on biota; N2 and PEL= threshold above which there is a higher probability of adverse effects on biota; n.a. = not determined.

Table 3.3 - Absolute (N) and relative (AR) abundances and ecological indices of foraminifer species recorded in the Una River estuarine channel during the first (I - Nov/2011) and second (II - Apr/2012) sampling campaigns

Una river canal	UI1	UI2	UI3	UI4	UI5	UI6	UI7	UI8	UI9	UI10	N	AR	UII1	UII2	UII3	UII4	UII5	UII6	UII7	UII8	UII9	UII10	N	AR
Ammonia beccarii	6	2	0	0	0	0	0	0	0	0	8	38,1	2	0	7	10	0	0	0	0	0	0	19	19,8
Ammonia tepida	0	0	0	0	0	1	0	0	0	0	1	4,8	0	0	0	0	0	0	0	0	0	0	0	0
Ammonia parkinsoniana	0	0	0	0	0	0	0	0	0	0	0	0	0	0	1	0	0	0	0	0	0	0	1	1
Cibicides sp	0	0	0	0	0	0	0	0	0	0	0	0	2	0	0	0	0	0	0	0	0	0	2	2,1
Elphidium excavatum	1	1	0	0	0	0	0	0	0	0	2	9,5	2	0	2	1	0	0	0	0	0	0	5	5,2
Haplophragmoides wilberti	6	3	0	0	0	0	0	0	0	0	9	42,9	8	2	4	9	0	0	0	0	0	0	23	24
Miliammina fusca	0	0	0	0	0	0	0	0	0	0	0	0	0	0	1	3	0	0	0	0	0	0	4	4,2
Quinqueloculina seminula	0	0	0	0	0	0	0	0	0	0	0	0	0	0	1	0	0	0	0	0	0	0	1	1
Spirillina sp	0	0	0	0	0	0	0	0	0	0	0	0	0	0	1	0	0	0	0	0	0	0	1	1
Textularia aglutinans	0	0	0	0	0	0	0	0	0	0	0	0	0	1	1	0	0	0	0	0	0	0	2	2,1
Trochammina inflata	0	0	0	0	0	0	0	0	0	0	0	0	4	2	0	30	0	0	0	0	0	0	36	37,5
Trochammina squamata	0	0	1	0	0	0	0	0	0	0	1	4,8	1	0	1	0	0	0	0	0	0	0	2	2,1
N per point	13	6	1	0	0	1	0	0	0	0	21	100	19	5	19	53	0	0	0	0	0	0	96	100
No. of species	3	3	1	0	0	1	0	0	0	0	-	-	6	3	9	5	0	0	0	0	0	0	-	-
Wealth (Margalef)	0,8	1,1	-	-	-	-	-	-	-	-	-	-	1,7	1,2	2,7	1	-	-	-	-	-	-	-	-
Evenness (Pielou)	0,8	0,9	-	-	-	-	-	-	-	-	-	-	0,9	1	0,9	0,7	-	-	-	-	-	-	-	-
Diversity (S-Wiener)	0,7	0,7	0	-	-	0	-	-	-	-	-	-	1,2	0,7	1,4	1,1	-	-	-	-	-	-	-	-

Legend: UI1 = Una River from the first campaign at point 1; N = Total; RA = Relative Abundance; Diversity (S - Wiener) = Diversity (Shannon-Wiener)

Figure 3.3 - Percentages of the types of coloration and wear of foraminifera in the estuarine channel of the Una River, referring to the Nov/2011 and Apr/2012 campaigns

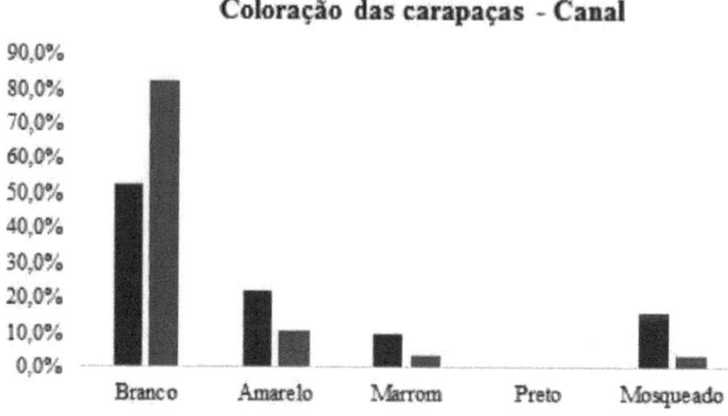

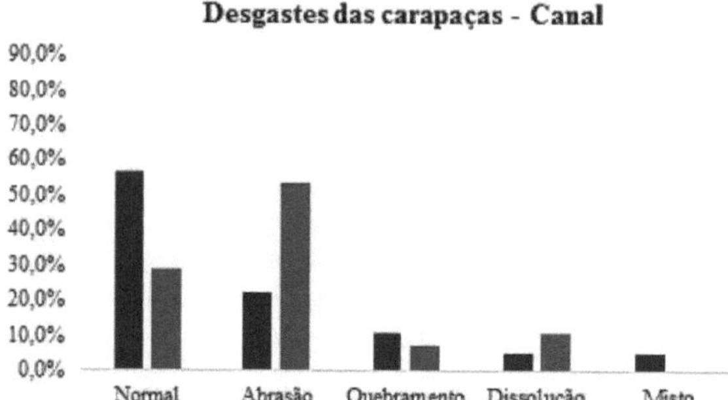

Legenda:
■ Periodo Chuvoso
■ Periodo Seco

1.1.5 Foraminifera distribution

In the principal component analysis (Figure 3.4), the main factor in the rainy season was the coarse sand values, indicating the existence of high hydrodynamic energy in the estuarine channel. In this period, the distribution of live individuals was correlated with the Fe>Pb>Cr>NT>P>Cu >Mn>Ni contents and showed a negative correlation with MO>OD, i.e. they were present in places where there was high consumption of dissolved oxygen and organic matter and nitrification and chemical oxidation of metal ions such as iron and manganese (CELINO et al., 2014).

On the other hand, in the dry season, the main factor was lead levels, which were negatively correlated with AG>Mn>T (Figure 3.4). In this campaign, the number of living organisms was related to the OD>MO>AF levels (Figure 3.4), *since* the reduction in rainfall and, consequently, hydrodynamic energy enabled the precipitation of fine sediment and the concentration of nutrients, thus favoring primary productivity in these sites (CELINO et al., 2014).

Figure 3.4 - Graphical representation of the Principal Component Analysis of biotic, abiotic, metal and nutrient parameters of the Una River estuarine channel, concerning the campaigns carried out in Nov/2011 (rainy season) and Apr/2012 (dry season).

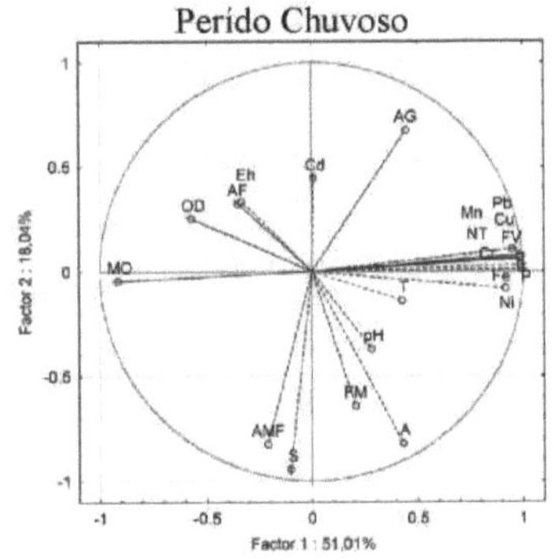

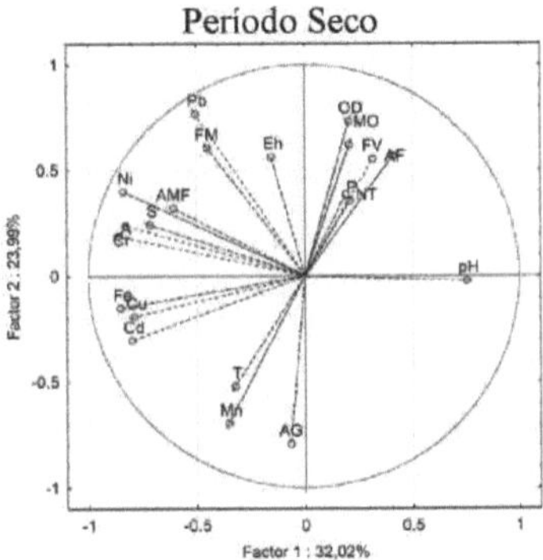

Note: Metals have been symbolized according to the periodic table, but FV = live foraminifera; FM = dead foraminifera; T = temperature; OD = dissolved oxygen; pH = hydrogen potential; Eh = oxidation-reduction potential; AG = coarse sand; AF = fine sand; AMF = very fine sand; S = silt; A = clay; NT = total nitrogen; MO = organic matter.

The distribution of dead individuals, in turn, showed a negative correlation with the levels of coarse sand in both campaigns (Figure 3.4). Bearing in mind that the metal levels in the sediment are below the limits adverse to biota (Table 3.2), corroborated even by the absence of anomalies in the tests, the presence of dead foraminifera is attributed to the influence of hydrodynamic energy, since, after the death of these organisms, the tests are usually deposited in lower energy environments, where, consequently, fine-grained sediments predominate (MACHADO et al., 2012).

3.3.6 Una River mangrove swamp: physico-chemical parameters

With regard to the interstitial waters in the mangrove zones of the Una River, the temperature values were also lower in the first campaign (24.0°C to 25.0°C) than in the second (28.1°C to 29.2°C) (Figure 3.5).

The sediment of the Una River mangrove was slightly acidic to neutral in both samples (6.6 to 7.3 in the first campaign and 6.0 to 7.1 in the second campaign) (Figure 3.5), which is within the expected range for mangrove areas since the decomposition of mangrove leaves causes the soil to show pH oscillations between 4.8 and 8.8 (BERRÊDO et al., 2008).

The Eh values of the sediment were negative at points 1 and 2 (-24.0 mV and -16.0 mV, respectively), characterizing them as a reducing environment, but from point 3 onwards the values were positive, showing the characteristics of an oxidizing environment at these sites during the first campaign. However, in the second campaign, the values were negative from points 2 to 4, giving them a predominantly reducing character, and positive at points 1, 5 and 6 (Figure 3.5).

31

3.3.7 Sediment analysis

There was a predominance of silt and very fine sand in both sampling periods, corroborating the low hydrodynamic energy depositional environment characteristic of mangrove areas (Figure 3.4).

The high levels of assimilable phosphorus (132.5 to 360.0 mg/L in the rainy season and 42.1 to 202.4 mg/L in the dry period - Table 3.4), especially at the points furthest upstream, suggest the influence of anthropogenic sources, such as domestic effluents or shrimp farming, emitting waste rich in phosphorus which, in turn, is easily adsorbed to the fine sediment of the region (MARTINS et al., 2011), so during the rainy season, the increased flow of the river remobilizes the phosphorus in the water column and transports it to the mangroves, further increasing its concentration.

Although all the collection points were located close to the municipality of Una, the total nitrogen values were considered low (0.2 mg/L to 0.3 mg/L in the first campaign and 0.1 to 0.3 mg/L in the second - Table 3.4), suggesting that it does not come from anthropogenic actions, but from natural sources based on primary biological production in the aquatic system, such as assimilation processes or consumption by phytoplankton (REEF et al., 2010).

Figure 3.5 - Values of the physicochemical parameters temperature, hydrogen potential and oxidation potential of the interstitial water and percentages of the granulometric fractions of the sediment in the mangrove zone of the Una river estuary, relative to the Nov/2011 (rainy season) and Apr/2012 (dry season) campaigns.

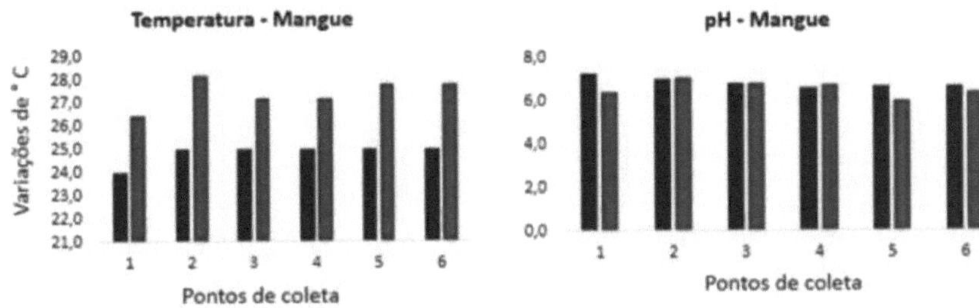

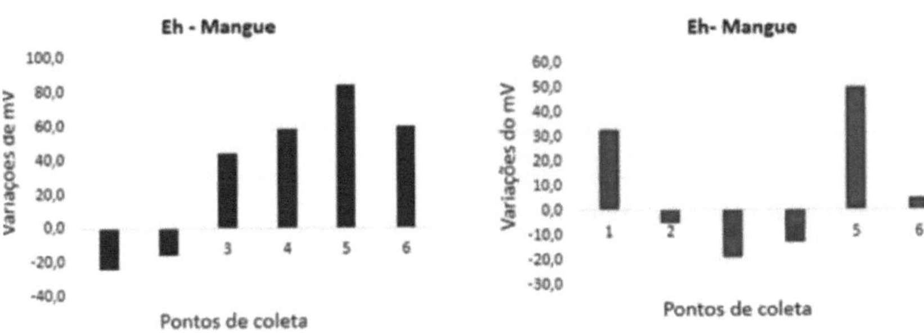

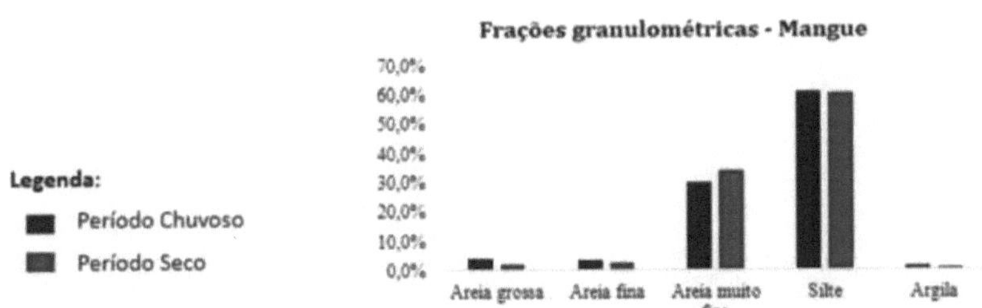

Table 3.4 - Physico-chemical, granulometric and nutrient data from the mangrove zone of the Una River for the first (I - Nov/2011) and second (II - Apr/2012) sampling campaigns

Samples	Water parameter	Sediment analysis								
	T (°C)	PH	Eh (mV)	GA (%)	PA (%)	MFA (%)	S (%)	A (%)	NT (mg/L)	P (mg/L)
MUI1	24,0	7,3	-24	2,7	6,2	37,2	52,7	1,2	0,2	270
MUI2	25,0	7,0	-16	2,2	0,0	20,9	75,6	1,4	0,2	132,5
MUI3	25,0	6,8	44	0,9	23,8	38,1	36,7	0,5	0,2	275

Samples	Sal	T	pH	Eh	AG	AF	AMF	S	A	P	NT
MUI4	25,0		6,6	58	6,6	0,0	23,6	67,8	2,0	0,3	345
MUI5	25,0		6,7	84	5,5	0,0	18,7	73,9	2,0	0,3	330
MUI6	25,0		6,7	60	17,3	0,0	19,4	61,6	1,7	0,3	360
Average	14,9		4,11	28,6	3,52	3,0	15,79	36,83	0,88	0,15	171,25
MUII1	26,4		6,4	33	0,7	9,1	39,4	50,2	0,6	0,1	42,1
MUII2	28,2		7,1	-5	1,0	0,1	32,1	65,4	1,4	0,0	98,7
MUII3	27,2		6,8	-19	0,6	3,3	43,6	51,8	0,8	0,3	184,8
MUII4	27,2		6,8	-13	2,2	0,0	28,5	67,8	1,4	0,2	202,4
MUII5	27,8		6,0	50	7,0	2,3	32,8	57,1	0,8	0,3	168,2
MUII6	27,8		6,5	5	0,1	0,0	27,8	70,6	1,6	0,3	137,6
Average	16,46		3,96	12,5	1,16	1,48	20,42	36,29	0,66	0,12	83,38

Legend: MUI1 = Una River mangrove swamp from the first campaign at point 1; Sal = Salinity; T = Temperature; pH = Hydrogen Potential; Eh = Oxirreduction Potential; AG = Coarse Sand; AF = Fine Sand; AMF = Very Fine Sand; S = Silt; A = Clay; P = Phosphorus; NT= Total Nitrogen; CONAMA = National Environmental Council.

Comparing the results obtained for the trace metal concentrations (Table 3.5) with the limits established by CONAMA and CEQG, it can be seen that only cadmium and lead showed values above the reference limits at some points, which may be related to the drainage of fertilizers and chemical compounds used in the agricultural development of the upstream region (ZOURARAH et al., 2009). In addition, all the elements showed higher levels in the dry season due to the reduction in flow and the consequent increase in concentrations in the water, allowing them to be deposited in the sediment, especially at the upstream points (Table 3.5).

Table 3.5 - Concentrations of metals (in mg. Kg⁻) in the sediment of the mangrove zone of the Una River, concerning the first (I - Nov/2011) and second (II - Apr/2012) sampling campaigns and reference values of CONAMA Resolution No. 454/2012 (Brazil, 2012) and the *Canadian Council of Ministers of the Environment* (CCME, 1998).

Samples	Ni	Mn	Fe	Cr	Zn	Cu	Pb	Cd
MUI1	9,68	198,15	15861,33	28,78	34,76	7,74	29,27	0,69
MUI2	6,36	60,70	14672,06	21,27	22,83	6,89	18,11	0,58
MUI3	8,17	195,60	16543,55	24,58	30,72	7,12	24,36	0,67
MUI4	10,1	246,06	19410,38	30,77	37,24	9,36	30,05	**0,86**
MUI5	9,64	232,49	21285,32	32,29	39,21	10,05	**31,24**	1,02
MUI6	10,87	316,34	21287,6	34,95	42,26	10,93	**34,53**	1,03
Average	9,14	208,22	18176,71	28,77	34,5	8,68	27,93	0,81
MUII1	4,12	52,14	7983,92	11,60	21,02	3,92	11,90	0,29
MUII2	8,74	131,30	17558,99	26,85	39,06	11,49	26,49	**0,73**
MUII3	12,22	245,63	22944,76	37,93	54,86	12,41	**36,33**	**1,12**
MUII4	13,46	334,12	23596,87	40,65	54,07	12,80	**42,11**	**1,20**
MUII5	7,68	136,02	18659,74	26,77	38,17	9,37	28,94	**0,82**
MUII6	12,88	430,51	24615,40	42,12	55,75	13,86	**41,53**	**1,25**
Average	9,85	221,62	19226,61	30,99	43,82	10,64	31,22	0,90
LD	0,00415	0,00111	0,02586	0,01632	0,00651	0,00336	1,49	0,00095
CONAMA N1	20,90	n.a.	n.a.	81	150	34	46,70	1,20
CONAMA N2	51,60	n.a.	n.a.	370	410	270	218	7,20
CEQG ISQG	n.a.	n.a.	n.a.	52,30	124	18,7	30,20	0,70
CEQG PEL	n.a.	n.a.	n.a.	160	271	108	112	4,20

Legend: LD = Limit of Detection; N1 and ISQG= threshold below which there is a lower probability of adverse effects on biota; N2 and PEL= threshold above which there is a higher probability of adverse effects on biota; n.a. = not determined.

3.3.8 Foraminifera fauna

In the first campaign (Nov/2011), 483 individuals were obtained, of which only 10 were alive. No fish were found at point 6 and, despite the high concentrations of cadmium and lead, no anomalous fish were recorded at any of the points. In this sampling, 13 species were identified, with *Miliammina fusca* (28.4%), *A. beccarii* (26.3%), *H. wilberti* (21.5%), *T. inflata* (15.1%) and *Trochammina squamata* (5.4%) considered the main ones (Table 3.6).

In the second campaign, 86 foraminifera were obtained, of which only 9 were alive. No foreheads were recorded at point 5, nor were there any anomalous individuals. Twelve species were identified, of which *H. wilberti* (27.9%), *Ammonia tepida* (25.6%), *T. inflata* (22.1%) and *E. excavatum* (8.1%) were considered the main ones (Table 3.6).

Haploprhagmoides and *Trochammina* species are routinely found in association in mangrove swamps due to the presence of fine sediment rich in organic matter (GÓMEZ; BERNAL, 2013; MACHADO et al, 2012), while *M. fusca* is recorded preferentially in areas with low salinity, such as near tributaries (DEBENAY et al., 1996; DEBENAY; GUILLOU 2002), so its occurrence in the rainy season is due to the increased fluvial influence on the mangrove. On the other hand, A. *beccarii, A. tepida* and *E. excavatum* are opportunistic species, so their presence in the studied area was favored by the supply of nutrients (SEMENSATTO-JR et al., 2009) and their tolerance to high concentrations of trace metals (MARTINS et al., 2013).

The richness values were higher in the second sampling (0.6 to 1.9 in the rainy season and 1.0 to 2.1 in the dry season - Table 3.6), but in both campaigns, the evenness indices were above 0.5, indicating a lack of dominance (CLARKE; WARWICK, 2001). Diversity was slightly lower in the dry season (1.1 to 1.4 in the first campaign and 0.8 to 1.4 in the second) (Table 3.6), but can be considered high when compared to the data obtained by Semensatto-Jr et al (2009) in a study carried out in a mangrove environment located to the north of Ilha do Cardoso, Cananéia-Iguape Bay - SP (0.2 to 0.6).

3.3.9 Taphonomy of the foreheads

In both sampling periods, most of the tests were white/colored (61.40% in the first and 61.32% in the second) or yellow (21.77% and 21.81%). As for wear, normal tests predominated (44.35% and 44.24%), although the percentage of abrasion was significant (28.95% and 29.01%) (Figure 3.6).

As occurred in the estuarine channel, the dominance of white foreheads results from the rapid addition of new foreheads to the sediment (MACHADO et al., 2012), which is confirmed by the predominance of normal foreheads (Figure 3.6). However, there was also a high percentage of abraded foreheads, which occurs in high-energy conditions, but these specimens were of *A. beccarii,* which also lives in the estuarine channel of the Una River (Tables 3.3 and 3.6), so the abraded foreheads were transported by the tides into the mangrove.

Figure 3.6 - Percentages of the types of coloration and wear of the foraminifer species in the mangrove zone of the Una River, referring to the Nov/2011 and Apr/2012 campaigns.

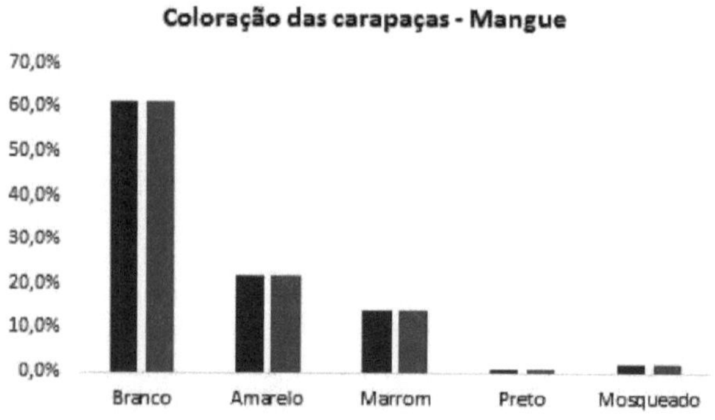

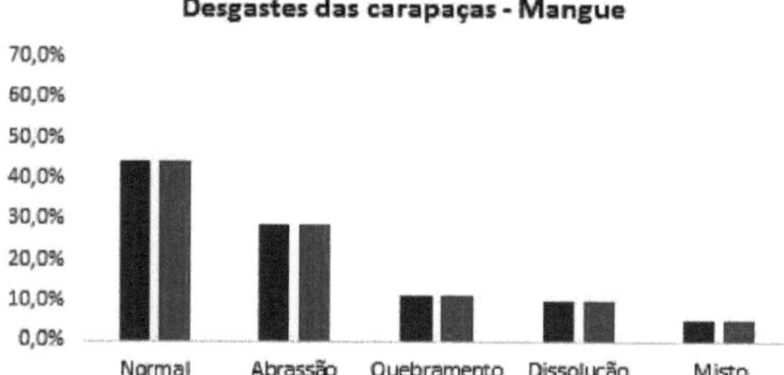

Legenda:
■ Período Chuvoso
■ Período Seco

Table 3.6 - Absolute (N) and relative (AR) abundances and ecological indices of foraminifer species recorded in the mangrove zone of the Una river estuary during the first (I - Nov/2011) and second (II - Apr/2012) sampling campaigns

Una river mangrove	UI1	UI2	UI3	UI4	UI5	UI6	N	AR	UII1	UII2	UII3	UII4	UII5	UII6	N	AR
Ammonia beccarii	5	0	91	21	10	0	127	**26,3**	0	0	1	0	0	0	1	1,2
Ammonia tepida	0	0	0	0	0	0	0	0	0	0	5	12	0	5	22	**25,6**
Bolivina disformis	0	0	1	0	0	0	1	0,2	0	0	1	0	0	0	1	1,2
Bolivina pulchella	0	0	1	0	0	0	1	0,2	0	0	0	0	0	0	0	0
Bolivina striatula	0	2	0	0	0	0	2	0,4	0	0	0	0	0	0	0	0
Cibicides sp.	0	0	0	0	0	0	0	0	0	2	0	0	0	0	2	2,3
Elphidium bertelotti	0	0	1	0	0	0	1	0,2	0	0	0	0	0	0	0	0
Elphidium excavatum	0	0	5	0	0	0	5	1	0	5	2	0	0	0	7	**8,1**
Elphidium poeyanum	0	0	0	0	0	0	0	0	0	1	0	0	0	0	1	1,2

Elphidium sagrum	0	0	1	0	0	0	1	**0,2**	0	0	0	0	0	0	0	0
Haplophragmoides wilberti	10	0	29	4	61	0	104	**21,5**	1	3	5	8	0	7	24	**27,9**
Miliammina fusca	40	0	15	22	60	0	137	**28,4**	0	0	0	2	0	1	3	3,5
Nonion grateloupi	0	0	3	0	0	0	3	0,6	0	0	0	0	0	0	0	0
Quinqueloculina lamarckiana	0	0	0	0	0	0	0	0	0	0	1	0	0	0	1	1,2
Saccammina sphaerica	0	0	2	0	0	0	2	0,4	0	0	0	0	0	0	0	0
Textularia aglutinans	0	0	0	0	0	0	0	0	0	2	0	0	0	0	2	2,3
Trochammina inflata	10	0	48	15	0	0	73	**15,1**	0	3	14	1	0	1	19	**22,1**
Trochammina squamata	6	0	0	0	20	0	26	**5,4**	0	1	2	0	0	0	3	3,5
N per point	71	2	197	62	151	0	**483**	100	1	17	31	23	0	14	**86**	100
No. of species	5	1	11	4	4	0	-	-	1	7	8	4	0	4	-	-
Wealth (Margalef)	0,9	0	1,9	0,7	0,6	-	-	-	-	2,1	2	1	-	1,1	-	-
Evenness (Pielou)	0,8	-	0,6	0,9	0,9	-	-	-	-	0,9	0,8	0,8	-	0,8	-	-
Diversity (S-Wiener)	1,2	0	1,4	1,2	1,1	-	-	-	0	1,4	1,4	0,9	-	0,9	-	-

Legend: UI1 = Una River from the first campaign at point 1; N = Total; RA = Relative Abundance; Diversity (S - Wiener) = Diversity (Shannon-Wiener)

3.3.10 Distribution of foraminifera

In the first campaign, the main factor was the content of very fine sand, followed by fine sand. The distribution of living individuals did not show a significant correlation with any of the parameters analyzed (Figure 3.7), suggesting that it was influenced by salinity values, which may be corroborated by the predominance of *M. fusca,* but this parameter was not measured during sampling. Dead individuals, on the other hand, were related to pH>AMF due to their preferential deposition in conditions of low hydrodynamic energy.

In the second sampling, the main factor was the total nitrogen content, followed by AG>Eh>P, with the number of live foraminifera being negatively related to these factors and positively related to pH>FM (Figure 3.7). According to Sarawast et al. (2015), the seasonal flow of freshwater in estuaries affects pH, causing its values to decrease along with salinity, which has adverse effects on the calcification and reproduction of foraminifera. Therefore, in this study, the association between the tests and pH is due to the increase in salinity during the dry season.

Figure 3.7 - Graphical representation of the Principal Component Analysis, based on biotic, abiotic, metal and nutrient parameters in the sediment of the Una River Mangrove Zone, concerning the campaigns carried out in Nov/2011 and Apr/2012

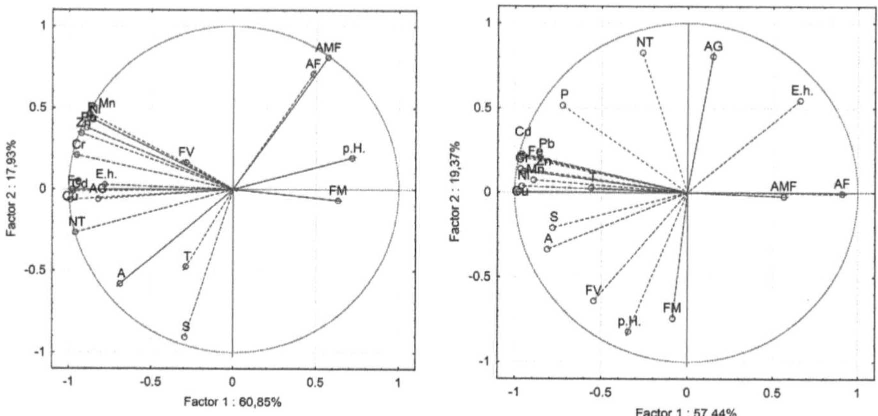

Note: Metals have been symbolized according to the periodic table, but FV = live foraminifera; FM = dead foraminifera; T = temperature; pH = hydrogen potential; Eh = oxidation-reduction potential; AG = coarse sand; AF = fine sand; AFM = very fine sand; S = silt; A = clay; NT = total nitrogen.

3.4 CONCLUSIONS

The estuarine channel of the Una River is made up of well-oxygenated waters, but due to an anomaly in the region's rainfall regime, the temperature values were lower in November 2011 than in April 2012.

The sediment was alkaline to slightly neutral and predominantly reductive during the rainy season, with coarse sand being the predominant granulometric fraction, although the percentages of the fine sand fraction were also high.

The concentration of metals in the sediment was below the limits adverse to biota, which was also corroborated by the absence of anomalies in the tests.

There was a predominance of recent samples of estuarine species, but the non-occurrence of foraminifera from point 3 in the rainy season and point 5 in the dry period was due to the low salinity values which made it impossible for even the most resistant species to survive. Despite this, there was no dominance of species, although the diversity values were considered low when compared to other estuaries.

Although the main factor in the rainy season was coarse sand, the distribution of living individuals was correlated with Fe>Pb>Cr>TN>P>Cu >Pb>Mn>Ni, and their negative correlation MO>OD indicates the presence of foraminifera in places with high consumption of dissolved oxygen and organic matter and nitrification and chemical oxidation of metallic ions such as iron and manganese. In the dry season, the main factor was lead levels, but the number of living organisms was related to the OD>MO>AG levels due to the increased precipitation of fine sediment and the concentration of nutrients.

On the other hand, the distribution of dead individuals showed a negative correlation with the levels of coarse sand in the two campaigns due to the influence of hydrodynamic energy, since the levels of metals in the sediment are below the limits adverse to biota.

In the mangrove swamp of the Una River, the high rainfall of the first campaign led to a reduction in the temperature and pH values of the sediment. In both sampling periods, the very fine sand and silt fractions were predominant.

The high concentrations of assimilable phosphorus in the mangrove zone come from anthropogenic sources, such as domestic effluents or carciniculture, as do the high levels of cadmium and lead which originate from agricultural drainage. In addition, all the elements showed higher values in the dry season due to the reduction in flow and, consequently, their dilution in the water.

There was a predominance of recent tests of estuarine species, but there was no dominance of species, and the diversity values were considered high when compared to other studies.

The levels of very fine sand and fine sand were the main factors during the rainy season, but the distribution of live foraminifera did not correlate with any of the parameters analyzed, suggesting that it was influenced by salinity levels. The distribution of dead individuals was correlated with pH>VS due to their preferential deposition in conditions of low hydrodynamic energy.

In the dry period, the main factor was the total nitrogen content, but foraminifera (live and dead) were related to pH due to the increase in salinity values.

3.5 REFERENCES

AZEVEDO, J.S.; BRAGA, E.S.; FAVARO, D.T.; PERRETTI, A.R.; REZENDE, C.E.; SOUZA, C.M.M. Total mercury in sediments and in Brazilian Ariidae catfish from two estuaries under different anthropogenic influence. **Marine Pollution Bulletin,** v.62, n.12, p.2724-2731, dec. 2011.

AGUIAR, P.C.B.; MOREAU, A.S.S.; FONTES, E.O. Impacts on the environmental dynamics of the municipality of Canavieiras (BA) with RESEX as a factor of influence. **Revista de Geografia, Meio Ambiente e Ensino,** v.2, n.1, p. 61-78, 1°Sem. 2011.

BAHIA (State). Secretariat for Territorial Development. **Territorial Plan for the Sustainable Development of the Territory: Southern Bahia Lowlands.** Bahia: PRONAT, 2010. p.136. Available at:

<http://sit.mda.gov.br/download/ptdrs/ptdrs_qua_territorio021.pdf>. Accessed on: April 22, 2015.

BOMFIM, C.S.; VILELA, C.G.; GUEDES, D.C. Benthic Foraminifera in Surface Sediments in the

Maricà Lagoon, Rio de Janeiro State. **Anuàrio do Instituto de Geociências,** v.33, n.1, p.9-19, 2010.

BRICKER, S.B.; LONGSTAFF, B.; DENNISON, W.; JONES, A.; BOICOURT, K.; WICKS, C.; WOERNER, J. Effects of nutrient enrichment in the nation's estuaries: a decade of change. **Harmful Algae,** v.8, n.1, p.21-32, dec. 2008.

BERRÊDO, J.F.; DA COSTA, M.L.; VILHENA, M.P.S.P.; SANTOS, J.T. Mineralogy and geochemistry of mangrove sediments from the Amazon coast: the example of the Marapanim river estuary (Parà). **Revista Brasileira de Geociências,** v.38, n.1, p.24-35, mar. 2008.

BRAZIL. Ministry of the Environment. National Environment Council, CONAMA. **CONAMA Resolution No. 357,** March 17, 2005. Available at: <http://www.mma.gov.br/port/conama/legiabre.cfm?codlegi=459>. Accessed on: March 13, 2014.

CELINO, J.J. et al. Geochemistry of surface water in the lower reaches of the Una, Pardo and Jequitinhonha rivers, southern Bahia. In: CELINO, J.J.; HADLICH, G.M.; QUEIROZ, A.F.S.; OLIVEIRA, O.M.C. (Orgs.). **Evaluation of coastal environments in the southern region of Bahia:** geochemistry, oil and society. Salvador: EDUFBA, 2014. p.63-76.

CRUZ, F.C. **Trace elements in mangrove substrate from the municipalities of Una, Canavieiras and Belmonte, Bahia.** 2012. 110f. Dissertation (Master's Degree in Geochemistry) - Institute of Geosciences, Federal University of Bahia, Salvador, 2012.

CARDOZO, L.S.; PORTO, M.F.; PIMENTEL, P.C.B.; RODRIGUES, J.S.; SCHIAVETTI, A.; CAMPIOLO, S. Discussion of the Board of the Extractive Reserve of Canavieiras, Bahia, Brazil: fisheries management to environmental management. **Revista da Gestao Costeira Integrada,** v.12, n.4, p.463-475, dec. 2012.

CARVALHO, P.V.; SANTOS, P.J.; BOTTER-CARVALHO, M.L. Assessing the severity of disturbance for intertidal and subtidal macrobenthos: the phylum-level meta-analysis approach in tropical estuarine sites of northeastern Brazil. **Marine Pollution Bulletin,** v. 60, n.6. p.87387, jun. 2010.

CLARKE, K.R.; WARWICK, R.M. **Change in marine communities:** An Approach to Statistical Analysis and Interpretation. 2nded. [s.l.]: Plymouth Marine Laboratory, 2001. 172p.

CARR, M.R. **Plymouth routines in multivariate ecological research user manual.** 2nd ed. [s.l.]: Plymouth Marine Laboratory, 1996.

CUNHA, L.H.DE O. **Reservas extrativistas:** uma alternativa de produção e conservaçao da biodiversidade. Sao Paulo: 1992. Available at:

<http://nupaub.fflch.usp.br/sites/nupaub.fflch.usp.br/files/color/resex.pdf>. Accessed on: Aug. 31, 2014.

DUQUET, M. **Ciências da vida:** glossàrio de ecologia fundamental. Porto: Porto Editora, 2007. 128p.

DONNICI, S.; SERANDREI-BARBERO, R.; BONARDI, M.; SPERLE, M. Benthic foraminifera as proxies of pollution: The case of Guanabara Bay (Brazil). **Marine Pollution Bulletin**, v.64, n.10, p.2015-2028, oct. 2012.

DE PAULA, F.C.F.; SILVA, D.M.L.; SOUZA, C.M. Typologias Hidroquimicas das Bacias Hidrogràficas do Leste da Bahia. **Revista Virtual de Quimica**, v.4, n.4, p.365-373, apr.2012.

DAUVIN, J.C. Paradox of estuarine quality: Benthic indicators and indices, consensus or debate for the future. **Marine Pullution Bulletin**, v.55, n. 1-6, p.271-281, 2007.

DEBENAY, J.P.; GUILLOU, J.J. Ecological Transitions Indicated by Foraminiferal Assemblages in Paralic Environments. **Estuaries**, v.25, n.6A, p. 1107-1120, dec. 2002.

DEBENAY, J.P.; BECK-EICHLER, B.; FERANNDEZ-GONZALEZ, M.; MATHIEU, R.; BONETTI, C.; DULEBA, W. Lês Foraminferes Paraliques Dês cotes D'Afrique et D'Amerique du sud de parte et D'autre de L'Atlantique: comparaison-discussion. Geology of Africa and the South Atlantic. **Actes Colloques Angers**, v.1994, p.463-471, Jun. 1996.

DAJOZ, R. **General ecology**. 2ed. Petrópolis: Vozes, 1983. 472p.

EICHLER, P.P.B.; EICHLER, B.B.; Gupta, B.S.; Rodrigues, A.R. Foraminifera as indicators of marine pollutant contamination on inner continental shelf of southern Brazil. **Marine Pollution Bulletin**, v.64, n.1, p.22-30, jan. 2012.

ESCOBAR, N.F.C.; CELINO, J.J.; NASCIMENTO, R.A. Metals in surface water, suspended particulate matter and bottom sediment in the lower reaches of the Una, Pardo and Jequitinhonha rivers. In: CELINO, J.J.; HADLICH, G.M.; QUEIROZ, A.F.S.; OLIVEIRA, O.M.C. (Orgs.). **Evaluation of coastal environments in the southern region of Bahia: geochemistry, oil and society**. Salvador: EDUFBA, p.77-98, 2014.

ESCOBAR, N.F.C. **Geochemistry of surface water and bottom sediment in the lower reaches of the Una, Pardo and Jequitinhonha rivers, Southern Bahia, Brazil**. 2013. 107f. Dissertation (Master's Degree in Geochemistry) - Institute of Geosciences, Federal University of Bahia, Salvador, 2013.

EIA - Environmental Impact Study. **Flood control system for the Una River Basin - Igarapeba Dam**. Recife, 2011. 47p. (Technical report).

FAGNANI, E.; GUIMARAES, J.R.; MOZETO, A.A.; FADINI, P.S. Acid volatile sulfides and simultaneously extracted metals in the assessment of freshwater sediments. **Quimica Nova,** v.34, n.9, p.1618-1628, maio. 2011.

FRONTALINI, F.; BUOSI, C.; DA PELO, S.; COCCIONI, R.; CHERCHI, A.; BUCCI, C. Benthic foraminifera as bio-indicators of trace element pollution in the heavily contaminated Santa Gilla lagoon (Cagliari, Italy). **Marine Pollution Bulletin,** v.58, n.6, p.858-877, jun. 2009.

FILHO, N.E.M. **Chemical characterization of organic matter in mangrove soil sediments and nutrient dynamics in surface and interstitial waters in the middle estuary of the Paciência River in Iguaiba - Paço do Lumiar (MA).** 2009. 162f Thesis (PhD in Chemistry) - Federal University of Paraiba, Joao Pessoa, 2009.

FRONTALINI, F.; COCCIONI, R. Benthic foraminiferal for heavy metal pollution monitoring: A case study from the central Adriatic Sea coast of Italy. **Estuarine Coastal and Shelf Science,** v.76, n.2, p.404-417, jan. 2008.

GÓMES, E.; BERNAL, G. Influence of the environmental characteristics of mangrove forests on recent benthic foraminifera in the Gulf of Urabà, Colombian Caribbean. **Marine Sciences,** v.39, n.1, p.69-82, 2013.

GONÇALVES, R.C.; COLONESE, J.; SILVA, M.; EGLER, S.; BIDONE, E.; CASTILHOS, Z.; POLIVANOV, H. Distribution of mercury, copper, lead, zinc and nickel in stream sediments of the Piabanha river basin, Rio de Janeiro State. **Geochimica Brasiliensis,** v.25, n.1, p.35-45, 2011.

GOMES, R.C.T. **Caracterizaçâo da fauna de foraminiferos da zona euhalina do Estuàrio do Rio Jacuipe - Camaçari-Ba.** 2010. 168f Dissertation (Master's in Geology) - Institute of Geosciences, Federal University of Bahia, Salvador, 2010.

IBGE - Brazilian Institute of Geography and Statistics. **Cidades@, Canavieiras - BA.** 2014. Available at: <http://www.cidades.ibge.gov.br/xtras/perfil.php?lang=&codmun=290630>. Accessed on: March 13, 2014.

LEGORBURU, I.; RODRÌGUEZ, J.G.; BORJA, A. MENCHACA, I.; SOLAUN, O.; VALENCIA, V.; GALPARSORO, I.; LARRETA, J. Source characterization and spatiotemporal evolution of the metal pollution in the sediments of the Basque estuaries (Bay of Biscay). **Marine Pollution Bulletin,** v.66, n.1-2, p.25-38, jan. 2013.

LOPES, C.M. **Individual and competitive adsorption of Cd, Cu, Ni and Zn in soils as a function of pH variation.** 2009. 102f. Dissertation (Master's Degree in Agronomy) - Luiz de Queiroz College of Agriculture, University of São Paulo, Piracicaba, 2009.

MARTINS, V.A. et al. Assessment of the health quality of Ria de Aveiro (Portugal): heavy metals and benthic foraminifera. **Marine Pollution Bulletin,** v.70, n.1-2, p.18-33, may. 2013.

MACHADO, A.J.; ARAÙJO, T.M.F.; De ARAÙJO, H.A.B.; FIGUEIREDO, S.M.C. Bathymetric and taphonomic analysis of the foraminiferal microfauna of the continental shelf and slope of the Municipality of Conde, Bahia. **Cadernos de Geociências,** v.9, n.2, p.157-172, nov. 2012.

MAIA, R.c.; cOuTINHO, R. Structural characteristics of mangrove forests in Brazilian estuaries: A comparative study. **Revista de Biologia Marina e Oceanografia,** v.47, n.1, p.8798, apr. 2012.

MARTINS, R.V.; FILHO, P.F.J.; ESCHRIQuE, S.A.; LACERDA, L.D. Anthropogenic sources and distribution of phosphorus in sediments from the Jaguaribe River estuary, NE, Brazil. **Brazilian Journal of Biology,** v.71, n.3, p.673-678, aug. 2011.

MARTINS, M.V.A.; DA SILVA, E.F.; SEQuEIRA, C.; ROCHA, F.; DuARTE, A.C. Evaluation of the ecological effects of heavy metals on the assemblages of benthic foraminifera of the canals of Aveiro (Portugal). **Estuarine, Coastal and Shelf Science (Print),** v.87, n.2, p.293-304, apr. 2010.

MuRRAY J. **Ecology and applications of benthic foraminifera.** Englan: Cambridge university Press, 2006. 426p.

ONOFRE, C.R.E.; CELINO, J.J.; NANO, R.M.W.; QuEIROZ, A.F.S. Bioavailability of trace metals in mangrove sediments from the northern portion of Todos os Santos Bay, Bahia, Brazil. **Revista de Biologia e Ciências da Terra,** v.7, n.2, p.65-82, 2Sem. 2007.

QuEIROZ, A.F.S.; OLIVEIRA, O.M.C. **Geoenvironmental diagnosis of mangrove areas and development of technological processes applicable to the remediation of these areas:** subsidies for an impact prevention program in areas with potential for oil activities in the southern coastal region of the State of Bahia (PETROTECMANGUE-BASUL). Salvador: EDuFBA, 2013. 148p. (Technical report)

REEF, R.; FELLER, I.C.; LOVELOCK, C.E. Nutrition of mangroves. **Tree Physiology,** v.30, p.1148-1160, 2010.

RODRIGUES, A.R.; MALUF, J.C.C.; BRAGA, E.S.; EICHLER, B.B. Recent benthic foraminiferal distribution and related environmental factors in Ezcurra Inlet (King George Island, Antarctica). **Antarctic Science,** v.22, n.4, p.343-360, 2010.

SARASWAT, R.; KOUTHANKER, M.; KURTARKAR, S.R.; NIGAM, R.; NAQVI, S.W.A.; LINSHY, V.N. Effect of salinity induced pH/alkalinity changes on benthic foraminifera: A laboratory culture experiment. **Estuarine, Coastal and Shelf Science,** v.153, p.96-107, feb. 2015.

SANTOS, L.C.M.; MATOS, H.R.; SCHAEFFER-NOVELLI, Y.; CUNHA-LIGNON, M.;

BITENCOURT, M.D.; KOEDAM, N.; DAHDOUH-GUEBAS, F.E. Anthroógenci activities on mangrove areas (Sao Francisco River Estuary, Brazil Northeast): A GIS-based analysis of CBERS and SPOT images to aid in local management. **Ocean and Coastal Management,** v.89, p.39-50, 2014.

SATYANARAYANA. B.; MULDER. S.; JAYATISSA. L.P.; DAHDOUH-GUEBAS. F. Are the mangroves in the Galle-Unawatuna area (Sri Lanka) at risk? A social-ecological approach involving local stakeholders for a better conservation policy. **Ocean and Coastal Management,** v.71, p.225-237, 2013.

SANTOS, L.C.; CUNHA-LIGNON, M.; SCHAEFFER-NOVELLI, Y.; CITRÓN-MORELO, G. Long-term effects of oil pollution in mangrove forests (Baixada Santista, Southeast Brazil) detected using a GIS-Based multitemporal analysis of aerial photographs. **Brazilian Journal of Oceanography,** v.60, p.161-172, 2012.

SEMENSATTO-JR., D.L.; FUNO, R.H.F.; DIAS-BRITO, D.; COELHO-JR., C. Foraminiferal ecological zonation along a Brazilian mangrove transect: Diversity, morphotypes and the influence of subaerial exposure time. **Revue de Micropaléontologie,** v.52, p.67-74, 2009.

TEODORO, A.C.; DULEBA, W.; GUBITOSO, S. Multidisciplinary study (geochemistry and foraminiferal associations) to characterize and evaluate anthropogenic interventions in Araçâ Bay. **Geosciences,** v.11, n.1, p.113-136, apr. 2011.

TEODORO, A.C.; DULEBA, W.; GUBITOSO, S.; PRADA, S.M.; LAMPARELLI, C.C.; BEVILACQUA, J.E. Analysis of foraminifera assemblages and sediment geochemical properties to characterize the environment near Araçâ and Saco da Capela domestic sewage submarine outfalls of Sao Sebastiao Channel, Sao Paulo State, Brazil. **Marine Pollution Bulletin,** v.60, p.536-553, 2010.

TEODORO, A.C.; DULEBA, W.; LAMPARELLI, C.C. Foraminiferal associations and textural composition of the region near the Cigarras domestic sewage submarine outfall, Sao Sebastiao Channel, SP, Brazil. **Pesquisa em Geociências,** v.38, p.467-475, 2009.

VEIGA, J.E. Indicadores de sustentabilidade. **Estudos Avançados,** v.24, n.68, p.39-52, 2010.

ZOURARAH, B.; MAANAM, M.; ROBIN, M. Sedimentary records of anthropogenic contribution to heavy metal content in Oum Er Bia estuary (Morocco). **Environ Chem Lett.,** v.7, p.67-78, 2009.

4 RESPONSES OF PARALIC FORAMINIFERA TO DIFFERENT SOURCES OF POLLUTANTS: A CASE STUDY OF THE CHANNEL AND MANGROVE ZONE OF THE RIO PARDO ESTUARIO, SOUTHERN LITERARAL OF BAHIA.

Summary

Foraminifera are widely used in studies of ancient and recent environmental variations, from natural and abiotic sources, highlighting their use as bioindicators of environmental stress. Thus, the species' associations with the characteristics of their foreheads were related to the concentrations of trace metals in the sediment of the mangrove zone of the estuarine channel of the Pardo River, on the southern coast of Bahia, with the aim of assessing whether the levels of these elements were affecting the microfauna. Two seasonal samplings were carried out (November/2011, April/2012) to collect 10 samples of bottom sediment in the channel and 6 samples of surface sediment near the *Avicennia* specimens in the mangrove, totaling 32 samples. In the first campaign, 467 foraminiferal tests were obtained from the canal region (15.8% of the specimens were collected alive and none of the tests were malformed) belonging to 8 species, including *Ammonia beccarii*; *Trochammina inflata; T. squamata*; *Haploprhagmoides wilberti*; *Elphidium excavatum* and *E. poeyanum*. In the second sampling, 96 testes (11.4% live; 0.0% anomalies) of 12 species were recorded, with a predominance of *A. tepida* ; *T. inflata* ; *H wilberti* ; *A. beccarii* ; *E. excavatum* and *T. squamata*. In the mangrove zone, 157 testes (9.5% live; 0.6% anomalous) of 8 species were obtained in the first campaign, with *H. wilberti, T. inflata,* T. *squamata* and *Quinqueloculinafusca standing* out. In the second campaign, only 43 tests were recorded (16.27% live; 0.0% anomalies) and 4 species: *H. wilberti*; *T. inflata*; *E. excavatum* and *A. beccarii*. The metal lead at points 2, 3, 5 and 6 (first campaign) and points 1, 5 and 6 (second campaign) was the only one to show high concentrations for the Pardo River Mangrove Zone, according to the Canadian Environmental Quality Guidelines (CCME, 1998). Thus, *A. beccarii* and *E. excavatum* are opportunistic species that are very common in impacted environments, but which are probably found in the region due to the supply of nutrients to the channel and mangrove zone. *H. wilberti* and *T. inflata, on the* other hand, are common in the mangrove, due to the predominance of sediments with a smaller grain size and consequent nutrient input, and *M. fusca* due to its preference for muddy sediments.

Keywords: Foraminifera, trace metals, Estuaries, Brazil

4.1 INTRODUCTION

The coastal zone stretches from the limits of the estuaries to the mangrove areas, corresponding to one of the controlling links in the flows and destinations of water and matter in the environment (KNOPPERS et al., 2006). Estuaries, in particular, have a very dynamic characteristic, resulting from the interaction and temporal variability resulting from the incidence of sea currents and river inputs (FÉLIX et al., 2012). Although this environment contributes significantly to primary oceanic production, as reactors in biogenic production at the land-sea interface, the flows of water and organic matter to the coastal zones have been altered due to anthropogenic actions, such as the management of drainage basins, the construction of river ports, fishing and recreational activities, as well as growing urbanization, caused by population growth (SANTOS et al., 2013).

Another environment impacted by anthropogenic actions are the mangrove zones, which from an environmental perspective are characterized as tropical ecosystems that contain plant communities typical of flooded environments with great variations in salinity, as well as acting as a nursery and nutrient source for a huge number of animal and plant species (SATYANARAYANA et al., 2013). In addition, the mangrove ecosystem is responsible for more than 10% of the dissolved organic carbon resulting from the processes of the biogeochemical cycle that flows from the land to the sea (STIEGLITZ et al., 2013). However, in the last two centuries, industrial development, population growth and the development of tourism have led to a disorderly occupation of coastal regions which can sometimes be justified by economic reasons (LOSADA, 2012).

Thus, there are several ways to study and characterize the conditions of this ecosystem, but the most effectively used is the one that combines biogeochemical analyses of aquatic organisms and geochemical analyses of sediments with studies of bioindicators (DAUVIN, 2007). Among these, paralic foraminifera stand out due to their high sensitivity to physical and chemical changes in the environment, in order to understand the natural dynamics of ecosystems and also to detect anthropogenic changes promoted in these environments (FRONTALINI et al., 2008; TEODORO et al., 2009; 2010; 2011; EICHLER et al., 2012).

The aim of this study was to describe the prevailing conditions in the estuarine channel and mangrove zone of the Pardo River, on the southern coast of the state of Bahia, by studying the characteristics of paralic foraminifera and the geochemical data of the sediment.

4.2 MATERIALS AND METHODS

The study area corresponds to the channel and mangrove area of the Pardo river estuary, whose banks are located in the municipality of Canavieiras, on the Dendê Coast, on the southern coast of the state of Bahia (Figure 4.1).

The Pardo river basin rises in the municipality of Rio Pardo de Minas, at an altitude of 750 m, in the state of Minas Gerais, and is 565 km long (XAVIER, 2009). It drains 27 municipalities in the southeast of the state of Bahia (PERH-BA, 2003), ending its course in the municipality of Canavieiras (15°41'S and 38°57'W), which has a population of 33,570 inhabitants and an area of 1,326.931 km^2 whose mangroves cover 0.07403 km^2 (5.4% of the municipality's total area) (IBGE, 2014).

In the estuarine region, alluvial soils are found, probably where the first cocoa plantations in Brazil were established, as well as coastal sands. There are also coastal islands, especially the islands of Passagem, due to the extent of the area, and Atalaia, whose beaches are the tourist attraction of the city of Canavieiras (AGUIAR et al., 2011).

The mangrove areas are cut by tidal channels and form an estuarine complex that connects the mouths of the Pardo and Jequitinhonha rivers. In addition to the tidal channels, the strong erosive action of the Pardo and Jequitinhonha rivers promotes the expansion of the river terraces, which have been occupied by the forest (SANTOS et al., 2002).

Figure 4.1 - Map showing the location of the municipality of Canavieiras and the sampling points in the canal and mangrove swamp

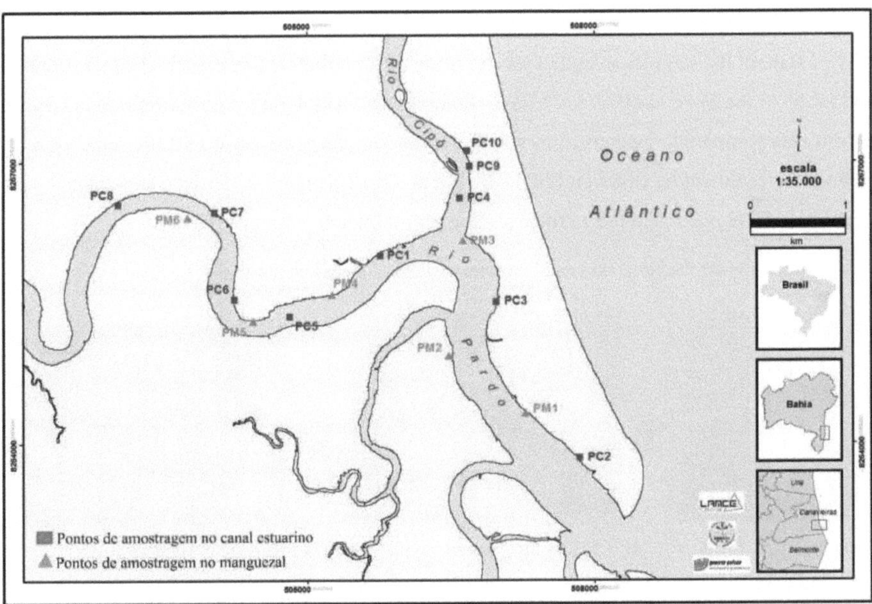

Source: QUEIROZ; OLIVEIRA, 2013.

4.2.1 Sampling procedure

Two collections were made - one during the dry season (25/11/2011) and the other at the start of the rainy season (23/04/2012) to obtain 32 samples of surface sediment at 10 points distributed in the estuarine channel of the Pardo River and at 6 points near specimens of *Avicennia* in mangrove areas in the municipality of Canavieiras (Figure 4.1).

The mangrove surface sediment samples were collected manually, but the estuarine channel sediment samples were obtained with the help of a van Veen-type dredger (Figure 2A). In both cases, two groups of samples were collected: one for the study of foraminifera and the other for granulometric and geochemical analysis of the sediment. The samples from the first group were packed in plastic bottles with screw caps and Rose Bengal dye was added to them to fix the live individuals captured at the time of collection (TEODORO et al., 2009), and they were kept refrigerated until the time of laboratory analysis (Figure 4.2B). The samples from the second group were stored in aluminum containers and frozen until they were processed in the laboratory.

The physico-chemical parameters (salinity, temperature and dissolved oxygen) of the estuarine channel water and mangrove interstitial water and (pH and Eh) of the channel and mangrove sediment were recorded using multiparameter probes.

4.2.2 Foraminifera fauna analysis

In the Laboratories of the Environmental Studies Center (NEA) (Institute of Geosciences/Federal University of Bahia), the samples intended for the study of foraminifera were washed under running water and taken to the oven at 60°C for 3 days (Figures 4.2C and 4.2D). After drying, about 3g of the sediment was poured into beakers, into which trichloroethylene was added for the separation of the tests from the sediment by flotation (DONNICI et al., 2012) (Figure 4.2E). The supernatant was then poured onto filter paper and left in the oven for approximately 5 minutes to dry.

Figure 4.2 - Procedures for the collection and laboratory analysis of sediment samples

A. Van Veen
B. Sediment stored with *rose bengal*
C. Rinsing in running water

D. Drying at 60 C° E. Flotation of the foreheads of F. Ideutification of foraminifera.
Ioiamiiiiferos.

In the Laboratory of the Foraminifera Study Group (LGEF), using a stereomicroscope, the specimens were removed from the filter paper and fixed with organic glue on microfossil slides. The foraminifera were identified on the basis of specialized literature and, during this procedure, information regarding their coloration and state of preservation was recorded (MACHADO et al., 2012) (Figure 4.2F). The occurrence of forehead anomalies was also recorded, when present.

4.2.3 Sediment analysis

The results of the analyses of the sediment from the channel and the mangrove swamp of the River Pardo were provided by NEA researchers, so the procedures for carrying out the analyses of granulometry, assimilable phosphorus, organic matter, organic carbon, total nitrogen and metals are described in Celino et al. (2014), Escobar et al. (2014) and Cruz (2012).

4.2.4 Statistical analysis

For the descriptive analysis of the foraminiferal fauna, the relative abundance and the relative frequency of occurrence were calculated (AB'SABER et al., 1997). Relative frequency (F) is defined as the ratio between the number of individuals in a category (n) and the total number of individuals in all categories (T), expressed as a percentage, namely

$$F = n.\frac{100}{T}$$

According to Dajoz (1983), the following classes were adopted: main (abundances >5%), accessory (4.9-1%) and trace (<1).

The frequency of occurrence is based on the number of occurrences of a category (p) in relation to the total number of samples (P) (AB'SABER et al., 1997).

$$C = p.\frac{100}{P}$$

Dajoz's (1983) classification was used to assess the frequency of occurrence, i.e. constant (occurrences >50%), ancillary (49-25%) and accidental (<24%).

In addition, the richness (Margalef index), evenness (Pielou index) and diversity (Shannon-Wiener index) indices were calculated using the Primer 6.0 program (CLARKE; WARWICK, 2001). In order to establish the relationship between foraminiferal distribution and environmental variables, the faunal, results were combined with those of the physico-chemical parameters and the granulometric and geochemical analyses to create a data matrix, from which principal component analyses were carried out using the Statistica 7.0 program (StatSoft, 2007).

4.   3RESULTS AND DISCUSSION

The results of the physicochemical parameters monitored in this study showed a rainfall anomaly in the months of the sampling, so that the rainfall was higher (332mm per month) during the first campaign (Nov 2011), which was characterized as a rainy season, while in the second sampling (Apr 2012) the rainfall rate was reduced (50mm monthly), characterizing it as a dry season (ESCOBAR, 2013), which certainly influenced the values of salinity, temperature and other parameters recorded during the collections. Therefore, for the purposes of this study only, the first campaign will be considered to have been carried out during the "rainy season" and the second during the "dry season".

4.3.1 Pardo River estuarine channel: physico-chemical parameters

The results of the physico-chemical parameters monitored were in line with the limits established in CONAMA Resolution No. 357 of March 17, 2005 (BRASIL, 2005) and, based on the salinity data, suggest that the water in the Pardo River estuary channel was classified as brackish in the first campaign and saline in the second (BRASIL, 2005) (Figure 4.3). Contrary to expectations, the salinity and temperature values were lower in the first (0.1 and 16.9 ups and 25.3 and 28.5 °C in Nov/2011) than in the second (1.7 and 31.0ups and 26.4 and 29.2°C in Apr/2012) sampling campaign (Figure 4.3). This is because during the sampling months there was an anomaly in the region's rainfall regime, so that it had a high rainfall index (332 mm per month) during the first campaign, behaving like a rainy period, while in the second sampling the rainfall rate was reduced (50 mm per month), characterizing it as a dry season (ESCOBAR, 2013). This behavior significantly influenced the data on salinity, temperature and the other parameters recorded during the collections, and the same was verified by Rolls (et al., 2012) in his work.

With regard to dissolved oxygen values, variations were observed (12.7 to 20.6 mg/L^{-1} , at points 3 and 9 in the first campaign, respectively, and 4.7 to 7.0 mg/L^{-1} , at points 3 and 7 in the second campaign, respectively) (Figure 4.3), since the low dissolved oxygen values recorded in the dry season are possibly related to the reduction in water volume and the rise in temperature in the month of sampling (AZEVEDO et al, 2011).

The pH values of the sediment were acidic to slightly neutral in both the first (4.2 at point 7 to 7.3 at point 1) and second campaigns (6.2 at point 2 to 7.2 at point 8) (Figure 4.3). In the first campaign, when we analyze the location of the collection points in the estuary, we see slightly acidic conditions from point 5 onwards, since the points upstream are more influenced by the humic acids from the decomposition of the mangrove's organic matter (BALDOTTO et al., 2011). In the second campaign, slightly acidic conditions were recorded at all points, indicating an increase in the concentration of acidic compounds due to the decrease in rainfall (Figure 4.3). For the Eh values of the sediment, in the first campaign negative values were obtained up to point 6 (-32.0 mV, -26.0 mV, -29.0 mV, -31.0 mV, -8.0 mV and -9.0 mV) and point 9 (-1.0 mV), classifying them as a reducing environment, while points 7 (143.0 mV), 8 (110.0 mV) and 10 (14.0 mV) were under oxidizing conditions. In the second campaign, only points 2 (-23.0 mV), 5 (-25.0 mV), 6 (-21.0 mV) and 7 (-16.0 mV) had negative Eh values (Figure 4.3). In fact, during the rainy season, increased rainfall causes redox potential to decrease (MESSIAS et al., 2013).

4.3.2 Sediment analysis

In both samples, fine sand was the predominant particle size, indicating a condition of low hydrodynamic energy in the channel (Figure 4.3).

In the dry season (April 2012), there was a reduction in the levels of assimilable phosphorus (0.2 mg/L at point 2 to 689.0 mg/L at point 6 in the first campaign and 2.5 mg/L at point 1 to 106.4 mg/L at point 4 in the second campaign), organic matter (0.1 mg/L at point 9 to 4,6 mg/L at point 6 in the first campaign and 0.1 mg/L at point 1 to 1.8 mg/L at point 6 in the second campaign) and organic carbon (0.1 mg/L at points 5, 9 and 10 to 2.7 mg/L at point 6 in the first campaign and 0.1 mg/L at points 3, 7 and 9 to 1.0 mg/L at point 6 in the second campaign). Only the total nitrogen values (0.1 mg/L at point 9 to 2.7 mg/L at point 6 in the first campaign and 0.2 mg/L at point 4 to 7.0 mg/L at point 6 in the second campaign) increased (Table 4.1). These results show that the concentrations of assimilable phosphorus, organic matter and organic carbon probably originate from agricultural drainage (diffuse sources) in the regions close to the upstream, a situation which becomes even more intense during the rainy season, while nitrogen results from urban and industrial effluents (point sources), which are more concentrated during the dry season (CARVALHO et al., 2010).

Chromium is generally used in stainless steel alloys, chromium plating, tanning, pigments, wood preservatives, organic syntheses and some types of fertilizers. The metal iron is one of the essential elements for life and its main source is the weathering of rocks and soils, so high concentrations are commonly found in estuarine regions. Nickel occurs naturally in the earth's crust, but its use in industry ranges from the production of nickel-copper alloys to inorganic pigments (FAGNANI et al., 2011). Cadmium, on the other hand, is considered a highly toxic non-essential metal that is usually present in high concentrations in sedimentary rocks and marine phosphates and in aquatic systems, so its presence is due to factors such as weathering, soil and bedrock erosion, direct atmospheric discharges from industries, landfill leaks and sites contaminated by the use of fertilizers in agriculture (FAGNANI et al., 2011). However, comparing the levels of the elements analyzed with the limits established by CONAMA Resolution No. 454/2012 (BRASIL, 2012) and by the *Canadian Council of Ministers of the Environment* (CCME, 1998), it was observed that the concentration of Ni, Cr, Cu, Pb and Cd in the sediment is low and, therefore, should not be adversely contaminating the biota of the Pardo River channel (Table 4.2). However, these standards do not set limits for Mn and Fe, which makes it impossible to assess the effect of these elements.

Figure 4.3 - Values of the physicochemical parameters, salinity, temperature, dissolved oxygen (D.O.) of the water and hydrogen potential (pH), oxidation potential (Eh) and granulometric fractions of the sediment of the estuarine channel of the Pardo River, relative to the campaigns of Nov/2011 (rainy season) and Apr/2012 (dry season).

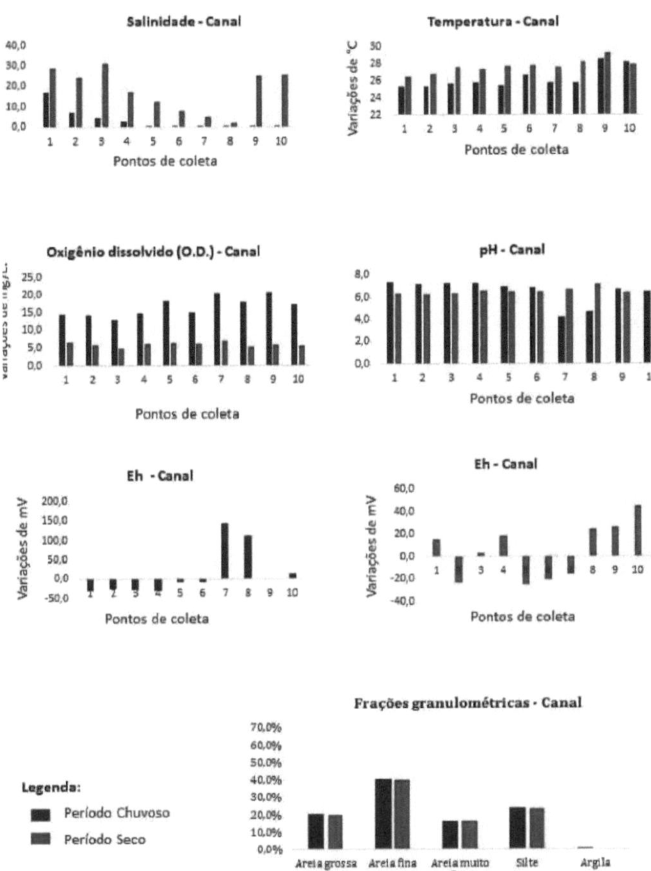

Table 4.1 - Physicochemical, granulometric and nutrient data of the water and sediment of the estuarine channel of the Pardo River for the first (I - Nov/2011) and second (II - Apr/2012) sampling campaigns and limits of CONAMA Resolution 357 of March 17, 2005 (Brazil, 2005).

Samples	Water parameters					Sediment Analysis							
	T (°C)	Salt (ups)	OD mg/L)⁻¹	(pH	Eh (mV)	AG(%)	PA (%)	MFA (%)	S (%)	A (%)	NT (mg/L)	P (mg/L)	MO (mg/L)
CPI1	25,3	16,9	14,5	7,3	-32	2,5	26,4	36,4	34,3	0,0	1,0	689	2,3
CPI2	25,3	7,1	14,2	7,2	-26	0,2	0,4	35,5	63,3	0,0	0,7	405000	3,3
CPI3	25,6	4,3	12,7	7,3	-29	22,8	61,9	9,5	5,8	0,0	1,8	437	3,9
CPI4	25,7	2,4	14,7	7,3	-31	57,6	16,5	9,4	16,4	0,0	0,3	309	1,6
CPI5	25,4	0,1	18,2	6,9	-8	18,0	70,4	7,5	4,1	0,0	0,0	274	0,2
CPI6	26,6	0,1	15,0	6,9	-9	0,5	0,0	23,5	73,8	0,0	2,7	0,2	4,6
CPI7	25,7	0,1	20,3	4,2	143	66,8	26,6	4,1	2,5	0,0	0,4	62,0	1,6
CPI8	25,7	0,1	18,0	4,7	110	65,4	24,6	3,4	5,8	0,0	0,3	58,2	0,6
CPI9	28,5	0,1	20,6	6,8	-1	18,0	70,5	7,6	4,0	0,0	0,1	18,9	0,1
CPI10	28,2	0,1	17,0	6,5	14	18,9	66,8	8,6	5,7	0,0	0,0	102,3	0,2
Average	26,2	3,1	16,5	6,5	40,3	27,1	36,4	14,6	21,6	0,0	0,7	40695,1	18,4
CPII1	26,4	28,6	6,8	6,3	15	27,3	62,7	5,6	4,5	0,0	0,5	9,9	0,0
CPII2	26,7	23,9	5,9	6,2	-23	37,6	53,8	4,8	3,8	0,0	1,7	2,5	0,7
CPII3	27,5	31,0	4,7	6,3	3	17,2	68,7	7,8	6,3	0,0	0,5	39,1	0,1

Samples	T	Sal	O.D.	pH	Eh	AG	AF	AMF	S	A	P	NT	
CPII4	27,3	16,7	6,2	6,5	18	47,2	46,1	3,5	3,2	0,0	0,2	106,4	0,0
CPII5	27,6	12,1	6,5	6,5	-25	12,9	56,0	13,4	17,5	0,3	3,8	49,1	0,9
CPII6	27,7	7,4	6,2	6,5	-21	5,7	46,3	18,0	29,6	0,4	7,0	79,6	1,8
CPII7	27,5	4,5	7,0	6,7	-16	6,3	0,0	24,8	67,4	1,5	1,0	22,0	0,3
CPII8	28,2	1,7	5,4	7,2	24	20,8	69,0	7,0	3,2	0,0	1,3	21,8	0,2
CPII9	29,2	25,0	5,8	6,5	26	16,3	61,1	8,8	11,9	2,0	1,5	36,8	0,2
CPII10	27,9	25,2	5,5	6,4	45	10,7	0,0	12,4	74,9	2,0	1,8	29,6	0,6
Average	27,6	17,61	6,0	6,5	4,6	20,2	46,4	10,6	22,2	0,6	1,9	39,7	4,8
CONAMA SWEET	-	<40,0	5	-	-	-	-	-	-	-	-	-	-
CONAMA SALINE	-	-	6	-	-	-	-	-	-	-	-	-	-
CONAMA SALOBRA	-	-	5	-	-	-	-	-	-	-	-	-	-

Legend: CPII = Pardo river channel from the first campaign; T = Temperature; Sal = Salinity; O.D. = Dissolved Oxygen; pH = Hydrogen Potential; Eh = Oxirreduction Potential; AG = Coarse Sand; AF = Fine Sand; AMF = Very Fine Sand; S = Silt; A = Clay; P = Phosphorus; NT = Total Nitrogen; CONAMA = National Environmental Council.

Table 4.2 - Concentrations of metals (in mg. Kg -1) in the sediment of the estuarine channel of the Pardo River, concerning the first (I - Nov/2011) and second (II - Apr/2012) sampling campaigns and reference values of CONAMA Resolution No. 454/2012 (Brazil, 2012) and *the Canadian Council of Ministers of the Environment* (CCME, 1998).

Samples	Ni	Mn	Fe	Cr	Cu	Pb	Cd
CPI1	11,59	158,51	11479,55	15,89	6,95	8,79	0,15
CPI2	7,53	67,69	6703,02	5,85	2,56	4,30	0,00
CPI3	15,71	172,73	18917,72	21,22	7,75	9,37	0,32
CPI4	4,54	40,92	4611,31	5,50	2,72	2,80	0,00
CPI5	3,04	8,35	1120,80	0,00	0,06	0,67	0,03
CPI6	13,35	317,26	26373,81	26,33	10,5	13,39	0,09
CPI7	0,01	68,87	6396,22	7,63	2,50	3,50	0,00
CPI8	0,01	19,86	3739,74	2,00	0,01	1,76	0,02
CPI9	0,01	18,55	2272,49	0,05	0,01	1,19	0,00
CPI10	3,51	16,98	1575,51	0,76	0,01	1,33	0,00
Average	5,93	88,972	83190,17	8,523	3,307	4,71	0,061
CPII1	0,12	8,27	689,12	0,50	0,16	0,01	0,40
CPII2	0,81	55,74	2220,96	2,57	0,50	1,24	0,40
CPII3	0,32	7,30	529,60	0,30	0,08	0,01	0,41
CPII4	0,40	10,84	1188,50	0,98	0,18	0,26	0,40
CPII5	3,87	58,17	7213,64	10,22	3,30	2,87	0,58
CPII6	1,52	40,57	3860,36	4,82	1,33	1,85	0,49
CPII7	0,42	7,53	1116,23	0,38	0,27	0,01	0,40
CPII8	0,27	3,80	956,52	1,05	0,15	0,10	0,37
CPII9	0,61	32,22	1600,69	1,21	0,18	0,34	0,44
CPII10	0,68	17,31	2320,74	2,18	0,47	0,46	0,44
Average	0,902	24,175	2,169,636	2,421	0,662	0,72	0,433
LD	0,0041	500,111	0,02586	0,01632	0,00336	1,49	0,00095
CONAMA N1	20,9	n.a.	n.a.	81	34	46,70	1,20
CONAMA N2	51,6	n.a.	n.a.	370	270	218	7,20
CEQG ISQG	n.a.	n.a.	n.a.	52,3	18,7	30,20	0,70
CEQG PEL	n.a.	n.a.	n.a.	160	108	112	4,21

Legend: LOD = Limit of Detection; N1 and ISQG= threshold below which there is a lower probability of adverse effects on biota; N2 and PEL= threshold above which there is a higher probability of adverse effects on biota; n.a. = not determined.

4.3.3 Foraminifera fauna

In the Rio Pardo channel, during the rainy season (Nov/2011), 467 samples were obtained, of which

74 were live organisms. At this station, 8 species were identified, with *Ammonia beccarii* (30.6%); *Elphidium excavatum* (9.4%); *E. poeyanum (5.1%)*; *Haploprhagmoides wilberti (12.2%)*; *Trochammina inflata (19.5%)* and *T. squamata (12.6%)* standing out as the main species (Table 4.3).

In the dry season, 96 individuals were obtained, of which 11 were alive. In this campaign, 12 species were identified, among which *A. beccarii* (7.1%); *A. tepida* (32.1%); *E. excavatum (7.1%)*; *H. wilberti (10.7%)*; *T. inflata* (17.9%) and *T. squamata (7.1%)*, were considered the main ones (Table 4.3).

The species belonging to the genera *Ammonia* and *Elphidium* are commonly found in seas and oceans, but can colonize estuarine environments, being recorded in brackish waters and in low-oxygen places (MURRAY, 2006). The species *H. wilberti* (point 1, point 3, point 4, point 5 and point 6); *T. squamata* (point 1, point 2, point 4, point 7 and point 10) and *T. inflata* (point 1, point 6 and point 10) are common in estuarine environments with low hydrodynamic energy and a tendency towards hyposalinity (BRUNO, 2013). The significant reduction and absence of foraminiferal tests is due to the high concentrations of cadmium, with the species *Elphidium excavatum*, *A. tepida* and *A. beccarii standing* out as the most tolerant to trace metal contamination (Tables 4.1 and 4.3) (GÓMES; BERNAL, 2013).

During the first campaign, the richness values ranged from (0.7 to 1.4 in the first campaign and 2.1 to 2.3 - in the second campaign), the evenness values ranged from (0.4 to 1.0 and 0.8 to 1.0), with only point 8 of the first campaign showing species dominance (values below 0.5 - CLARKE; WARWICK, 2001). The diversity values ranged from (0.8 to 1.7 and 1.0 to 1.3) (Table 3), which are considered low when compared to those obtained by Gomes (2010) in the Jacuipe River estuary (BA) (1.0 to 4.4), Teodoro (2009 and 2010) in the Sao Sebastiao Channel, SP (0.3 to 2.8) and Araça (2.4 to 2.9), also near the Sao Sebastiao-SP Channel.

The low diversity of foraminifer species found in the Pardo river channel is possibly related to the contamination of this site by trace metals and probably also to the influence of saltier waters during the second campaign (dry period).

Table 4.3 - Absolute (N) and relative (AR) abundances and ecological indices of the foraminifer species recorded in the estuarine channel of the Pardo River, relating to the first (I - Nov/2011) and second (II - Apr/2012) sampling campaigns

Pardo Canal	River	PI1	PI2	PI3	PI4	PI5	PI6	PI7	PI8	PI9	PI10	N	AR	PII1	PII2	PII3	PII4	PII5	PII6	PII7	PII8	PII9	PII10	N	AR
Ammonia beccarii		46	10	22	9	2	12	0	12	3	27	143	30,6	0	0	2	0	0	0	0	0	0	0	2	7,1
Ammonia tepida		0	0	0	0	0	0	0	0	0	0	0	0	0	0	9	0	0	0	0	0	0	0	9	32,1

First campaign:

Species											N	RA
Bolivina laevigata	0	0	0	0	0	0	0	0	0	0	0	0
Elphidium discoidale	0	0	0	0	0	0	0	0	0	0	0	0
Elphidium excavatum	26	0	11	0	0	7	0	0	0	0	44	9,4
Elphidium poeyanum	11	3	10	0	0	0	0	0	0	0	24	5,1
Haplophragmoides wilberti	20	0	13	2	2	20	0	0	0	0	57	12,2
Nonion grateloupi	3	0	0	0	0	0	0	0	0	0	3	0,6
Pyrgo nasuta	0	4	0	0	0	0	0	0	0	0	4	0,9
Quinqueloculina fusca	5	0	21	0	2	2	0	1	0	0	31	6,6
Spirillina decorata	0	0	9	0	0	0	0	0	0	0	9	1,9
Textularia aglutinans	0	0	0	0	2	0	0	0	0	0	2	0,4
Trochammina inflata	51	0	0	0	0	32	0	0	0	8	91	19,5
Triloculina sp.	0	0	0	0	0	0	0	0	0	0	0	0
Trochammina squamata	20	9	0	10	2	0	6	0	0	12	59	12,6
N per point	182	26	86	21	10	73	6	13	3	47	467	100
No. of species	8	4	6	3	4	5	1	2	1	3	-	-
Wealth (Margalef)	1,4	0,9	1,1	0,7	1,4	0,9	0	0,4	0	0,5	-	-
Evenness (Pielou)	0,9	0,9	1,0	0,9	1,0	0,8	*** *	0,4	*** *	0,9	-	-
Diversity (S-Wiener)	1,7	1,1	1,6	0,8	1,0	1,2	*** *	0,2	*** *	0,9	-	-

Second campaign:

Species											N	RA
Bolivina laevigata	0	0	1	0	0	0	0	0	0	0	1	3,6
Elphidium discoidale	0	0	0	0	0	0	0	0	0	0	0	0
Elphidium excavatum	0	2	0	0	0	0	0	0	0	0	2	7,1
Elphidium poeyanum	0	0	0	0	0	0	0	0	0	0	0	0
Haplophragmoides wilberti	0	1	2	0	0	0	0	0	0	0	3	10,7
Nonion grateloupi	0	0	0	0	0	0	0	0	0	0	0	0
Pyrgo nasuta	0	0	0	0	0	0	0	0	0	0	0	0
Quinqueloculina fusca	0	0	1	0	0	0	0	0	0	0	1	3,6
Spirillina decorata	0	0	1	0	0	0	0	0	0	0	1	3,6
Textularia aglutinans	0	1	0	0	0	0	0	0	0	0	1	3,6
Trochammina inflata	0	1	4	0	0	0	0	0	0	0	5	17,9
Triloculina sp.	0	0	1	0	0	0	0	0	0	0	1	3,6
Trochammina squamata	0	2	0	0	0	0	0	0	0	0	2	7,1
N per point	0	7	21	0	0	0	0	0	0	0	28	100
No. of species	0	5	8	0	0	0	0	0	0	0	-	-
Wealth (Margalef)	*** *	2,1	2,3	*** *	*** *	*** *	*** *	*** *	*** *	****	-	-
Evenness (Pielou)	*** *	1,0	0,8	*** *	*** *	*** *	*** *	*** *	*** *	****	-	-
Diversity (S-Wiener)	*** *	1,0	1,3	*** *	*** *	*** *	*** *	*** *	*** *	****	-	-

Legend: CPI1 = Pardo river channel from the first campaign at point 1; N = Total; RA = Relative Abundance; Diversity (S - Wiener) = Diversity (Shannon-Wiener).

4.3.4 Taphonomy of the foreheads

In both campaigns, the majority of the testes were white or colorless (52.25% in the first campaign and 82.14% in the second) or yellow (21.84% and 10.80%, respectively). In terms of wear, 56.75% of the foreheads were normal in the first campaign, but 53.60% showed signs of abrasion in the second campaign (Figure 4.4).

The predominance of white shells is indicative of a very rapid rate of deposition, with a lot of new material being added to the sediment, especially during the dry season. However, in this same sample it becomes predominant, indicating the existence of tidal transport of these shells (MORAES; MACHADO, 2003), which is corroborated by the increase in salinity values and the reduction in the number of shells of the species *H. wilbert*, *T. squamata* and *T. inflata* (Tables 4.1 and 4.3).

Figure 4.4 - Percentages of the types of coloration and wear of foraminifer species in the estuarine channel of the Pardo River, referring to the Nov/2011 and Apr/2012 campaigns

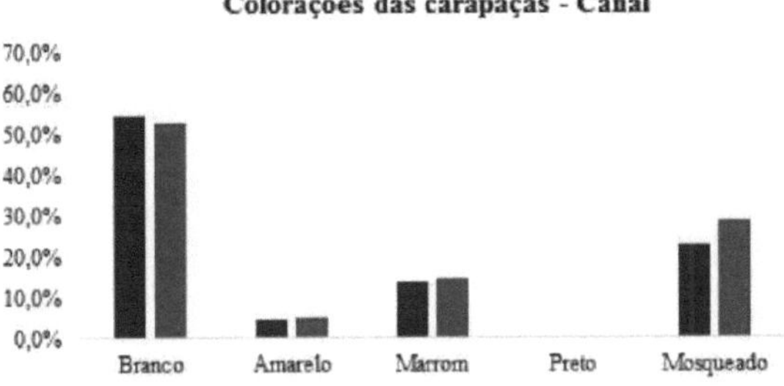

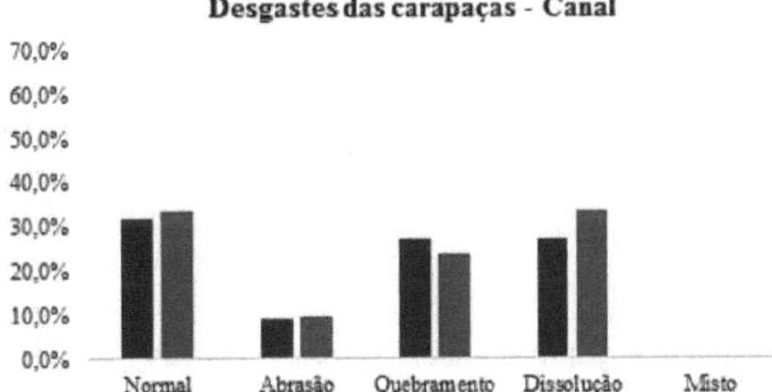

Legenda:
■ Periodo Chuvoso
■ Periodo Seco

4.3.5 Distribution of foraminifera

In the first campaign, the distribution of live individuals was positively correlated with dissolved oxygen (Figure 4.5), suggesting that these organisms prefer well-oxygenated environments. For the second campaign, the live ones showed a positive correlation with salinity, probably indicating that the increase in salinity favored the survival of the species (GÓMES; BERNAL, 2013).

The distribution of dead individuals showed a positive correlation in the first campaign with the elements (very fine sand>silt>phosphorus) (Figure 4.5) and in the second campaign (Figure 4.5) with fine sand, since, after the death of these organisms, the remains are usually deposited in environments with lower hydrodynamic energy, where, consequently, fine-grained sediments predominate, which, in turn, due to their small size, favor the deposition of phosphorus (EICHLER et al., 2007).

Figure 4.5 - Graphical representation of the Principal Component Analysis of biotic, abiotic, metal and nutrient parameters in the estuarine channel of the Pardo River, concerning the campaigns carried out in Nov/2011 and Apr/2012

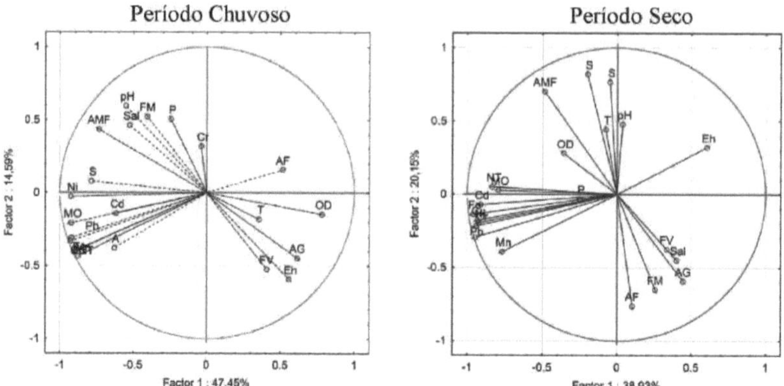

Note: Metals have been symbolized according to the periodic table, but FV = live foraminifera; FM = dead foraminifera; T = temperature; OD = dissolved oxygen; pH = hydrogen potential; Eh = oxidation-reduction potential; AG = coarse sand; AF = fine sand; AMF = very fine sand; S = silt; A = clay; NT = total nitrogen; MO = organic matter.

4.3.6 Pardo river estuary mangroves: physico-chemical parameters

With regard to the interstitial water in the mangrove zones, in the first campaign (considered the rainy season), the salinity values at points 4 to 6 were 0.0 ups, while at points 1, 2 and 3 the values recorded were 20.0 ups; 5.0ups and 1.0ups, respectively. In the second campaign (considered a dry period), salinity values ranged from 20.0 ups to 35.0 ups at the points (Figure 4.6).

The temperature during the first campaign ranged from 25.0°C to 25.8°C. While in the second sampling, the values varied between 25.3°C and 26.6°C (Figure 4.6).

In terms of pH, the Pardo River mangrove sediment was slightly acidic to neutral in both samples (6.3 to 7.2 in the first campaign and 6.4 to 7.1 in the second) (Figure 4.6), which is within the expected range for mangrove areas, since the decomposition of mangrove leaves causes the soil to fluctuate in pH between 4.8 and 8.8 (BERRÊDO et al., 2008).

The Eh values of the sediment were negative at points 1 and 2, characterizing them as a reducing environment, but from point 3 onwards the values were positive, showing an oxidizing environment at these sites during the first campaign. However, in the second campaign, the values were positive from point 1 to 6, except for point 5, showing a reducing condition for this point during the dry period (Figure 4.6).

4.3.7 Sediment analysis

There was a predominance of silt and very fine sand in both sampling periods, corroborating the low hydrodynamic energy depositional environment characteristic of mangrove areas (Figure 4.6).

The high levels of assimilable phosphorus (215.0 to 770.0 mg/L in the rainy season and 203.5 to 290.2mg/L in the dry season - Table 4.4), especially at the points furthest upstream, indicate a strong influence from anthropogenic sources, such as domestic effluents and shrimp farming activities, emitting phosphorus-rich waste which, in turn, is easily adsorbed to the region's fine sediment. In addition, during the rainy season, the increased flow of the river remobilizes phosphorus into the water column, further increasing its concentration (MARTINS et al., 2011).

Total nitrogen values were considered low (0.1mg/L to 0.4mg/L in the first campaign and 0.2 to 0.3mg/L in the second - Table 4.4), suggesting that it does not come from anthropogenic actions, but from natural sources based on the primary biological production of the aquatic system, such as assimilation processes or consumption by phytoplankton (REEF et al., 2010).

Figure 4.6 - Values of the physicochemical parameters, salinity, temperature, hydrogen potential and oxidation potential of the interstitial waters and percentages of the granulometric fractions of the sediment in the mangrove zone of the Pardo river estuary, relative to the campaigns of Nov/2011 (rainy season) and Apr/2012 (dry season).

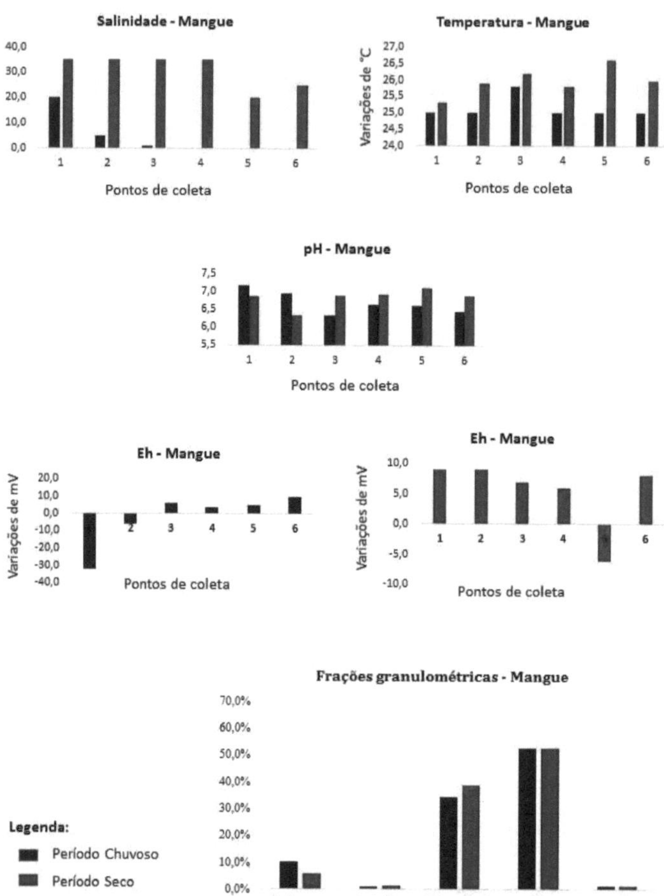

Table 4.4 - Physicochemical, granulometric and nutrient data of the water and sediment of the mangrove zone of the Pardo River for the first (I - Nov/2011) and second (II - Apr/2012) sampling campaigns and limits CONAMA Resolution 357 of March 17, 2005 (Brazil, 2005)

Samples	Water parameter				Sediment Analysis							
	Salt	T (°C)	PH	Eh (mV)	GA (%)	PA (%)	MFA (%)	S (%)	A (%)	NT (mg/L)	P (mg/L)
MPI1	20,0	25,0	7,2	-32	3,3	0,3	40,5	54,7	1,2	0,1	220
MPI2	5,0	25,0	7,0	-6	13,3	0,3	36,3	49,1	1,1	0,3	215
MPI3	1,0	25,8	6,3	6	13,1	0,0	21,8	63,4	1,8	0,3	390
MPI4	0,0	25,0	6,6	4	32,1	0,1	22,4	44,3	1,1	0,3	370
MPI5	0,0	25,0	6,6	5	32,1	0,1	22,7	44	1,2	0,3	770

Samples	Sal	T	pH	Eh	AG	AF	AMF	S	A	P	NT
MPI6	0,0	25,0	6,5	10	4,0	0,4	35,6	58,7	1,3	0,3	770
Average	2,6	15,08	4,02	6,3	9,79	0,12	17,93	31,42	0,77	0,16	273,5
MPII1	35,0	25,3	6,9	9	23,3	3,8	41,8	30,7	0,5	0,2	266,5
MPII2	35,0	25,9	6,4	9	6,1	2,0	44,3	46,9	0,7	0,0	208,1
MPII3	35,0	26,2	6,9	7	2,6	1,9	47,7	47,0	0,7	0,2	249,3
MPII4	35,0	25,8	6,9	6	1,2	1,8	44,1	51,8	1,1	0,2	257,3
MPII5	20,0	26,6	7,1	-6	0,3	0,0	25,6	71,5	2,6	0,0	203,5
MPII6	25,0	26,0	6,9	8	0,4	0,0	28,5	69,2	2,0	0,3	290,2
Average	18,5	15,58	4,11	4,5	3,39	0,95	23,2	31,71	0,76	0,09	147,49
CONAMA											
SWEET	< 5,0	<40,0	-	-	-	-	-	-	-	-	-
SALINE	> 30,0	-	-	-	-	-	-	-	-	-	-
SALOBRA	5,0 a 30,0	-	-	-	-	-	-	-	-	-	-

Legend: MPI1 = Pardo River Mangrove from the first campaign at point 1; Sal = Salinity; T = Temperature; pH = Hydrogen Potential; Eh = Oxirreduction Potential; AG = Coarse Sand; AF = Fine Sand; AMF = Very Fine Sand; S = Silt; A = Clay; P = Phosphorus; NT= Total Nitrogen; CONAMA = National Environment Council.

With regard to metals, comparing and analyzing the results obtained for trace metal concentrations (Table 4.5), with the limits established by CONAMA and CEQG, it was noted that only lead showed values above the reference limits in the first campaign (points 2, 3, 5 and 6) and in the second campaign (points 1, 5 and 6), which may be related to the drainage of fertilizers and chemical compounds used in agricultural development in the upstream region (ZOURARAH et al., 2009).

Table 4.5 - Concentrations of metals (mg. Kg -1) in the sediment of the mangrove zone of the Pardo River, concerning the first (I - Nov/2011) and second (II - Apr/2012) sampling campaigns and reference values of CONAMA Resolution No. 454/2012 (Brazil, 2012) and the *Canadian Council of Ministers of the Environment* (CCME, 1998).

Samples	Ni	Mn	Fe	Cr	Zn	Cu	Pb	Cd
MPI1	5,86	120,30	10441,38	14,50	19,87	4,98	19,76	0,35
MPI2	9,96	169,96	17955,67	27,60	37,41	10,59	**35,06**	0,70
MPI3	10,96	138,12	19188,10	31,10	41,25	12,64	**37,47**	0,87
MPI4	12,00	174,01	20200,16	31,90	45,73	12,95	46,15	0,89
MPI5	11,10	202,40	19866,29	29,80	43,70	12,12	**46,99**	0,92
MPI6	11,41	312,56	21353,40	33,60	45,87	13,52	**46,99**	1,03
Average	10,21	186,22	18167,5	28,08	38,97	11,13	38,73	0,79
MPII1	13,69	295,00	21377,69	37,20	57,15	14,04	**55,86**	0,95
MPII2	11,50	178,31	19020,26	31,00	49,14	12,10	40,71	0,80
MPII3	11,62	266,10	20037,75	31,60	50,02	12,60	42,65	0,88
MPII4	12,30	262,84	21038,57	34,70	54,09	13,38	45,39	0,99
MPII5	13,83	191,10	23378,24	41,00	64,94	28,26	**56,83**	1,15
MPII6	12,05	125,36	21542,62	35,10	58,05	17,56	**50,09**	0,95
Average	12,49	219,78	21065,9	31,1	55,56	16,32	48,58	0,95
LD	0,004	0,001	0,02586	0,01632	0,00651	0,00336	1,49	0,00095
CONAMA N1	20,9	n.a.	n.a.	81	150	34	46,7	1,2
CONAMA N2	51,6	n.a.	n.a.	370	410	270	218	7,2
CEQG ISQG	n.a.	n.a.	n.a.	52,3	124	18,7	30,2	0,7
CEQG PEL	n.a.	n.a.	n.a.	160	271	108	112	4,2

Legend: LOD = Limit of Detection; N1 and ISQG= threshold below which there is a lower probability of adverse effects on biota; N2 and PEL= threshold above which there is a higher probability of adverse effects on biota; n.a. = not determined.

4.3.8 Foraminifera fauna

For the first campaign (Nov. 2011), 157 foraminifera were obtained, of which 15 were alive and 0.63% of the foreheads were malformed. Eight species were identified, with *Haplophragmoides wilberti* (34.4%), *Trochammina inflata* (28.7%), *T. squamata* (14.7%) standing out as the main species and *Quinqueloculina fusca* (10.2%) (Table 4.6). The occurrence of these species indicates an environment with predominantly fine sedimentation and high organic content (MACHADO et al., 2012). In the second sampling (Apr/2012), 43 foraminifera were obtained, of which 7 were alive. Seven species were identified, most notably *H. wilberti* (30.2%), *T. inflata* (27.9%), *Elphidium excavatum* (23.3%) and *Ammonia beccarii* (7.0%). According to Debenay et al. (2000) the species *T. inflata* is associated with mangrove root systems, which in turn are rich in organic matter, a preferred environmental condition for the species *H. wilbert*. In the present study, during the second campaign, the highest percentages of testes obtained in the mangrove zone were from *H. wilbert* and *T. inflata,* corroborating the data from those authors. In addition, the significant reduction in the number of foraminifera in the second campaign is possibly due to the high concentrations of trace metals in the region, with *A. beccarii* and *E. excavatum* standing out as the species most tolerant to trace metal levels in the sediments (GÓMES; BERNAL, 2013).

The richness values were higher in the second sampling (0.6 to 1.4 in the rainy season and 1.1 to 1.4 in the dry season - Table 4.6), but in both campaigns, the evenness indices were above 0.5, indicating a lack of dominance (CLARKE; WARWICK, 2001). Diversity was slightly lower in the dry season (0.8 to 1.3 in the first campaign and 0.3 to 1.2 in the second) (Table 4.6). However, these values were considered high when compared to the data obtained by Semensatto-Jr (et al, 2009) in a study carried out in a mangrove environment located to the north of Ilha do Cardoso, in the Bay of Cananéia-Iguape-SP (0.2 to 0.6). Therefore, the high richness value in the dry season may be related to the tolerance of some species to contamination by trace metals (Tables 4.3 and 4.6).

Table 4.6 - Absolute (N) and relative (AR) abundances and ecological indices of the foraminifer species recorded in the mangrove zone of the Pardo river estuary during the first (I - Nov/2011) and second (II - Apr/2012) sampling campaigns

Pardo river mangrove	PI1	PI2	PI3	PI4	PI5	PI6	N	AR	PII1	PII2	PII3	PII4	PII5	PII6	N	AR
Ammonia beccarii	1	0	0	0	3	0	4	2,6	2	0	1	0	0	0	3	**7**
Bolivina laevigata	0	0	1	0	0	0	1	0,6	2	0	0	0	0	0	2	4,7
Elphidium discoidale	0	0	4	0	0	0	4	2,6	0	0	0	0	0	0	0	0
Elphidium excavatum	0	0	0	0	0	0	0	0	9	1	0	0	0	0	10	**23,3**
Elphidium	0	0	0	0	0	0	0	0	0	0	0	0	0	0	0	0

poeyanum																
Globigerina parchyderma	0	0	0	0	0	0	0	0	0	0	1	0	0	0	1	2,3
Globigerina trilobus	0	0	0	0	0	0	0	0	0	2	0	0	0	0	2	4,7
Haplophragmoides wilberti	2	19	19	14	0	0	54	**34,4**	11	0	0	0	1	1	13	**30,2**
Nonion grateloupi	1	0	1	0	0	0	2	1,3	0	0	0	0	0	0	0	0
Pyrgo nasuta	0	0	0	5	0	0	5	3,2	0	0	0	0	0	0	0	0
Quinqueloculina fusca	0	0	0	16	0	0	16	**10,2**	0	0	0	0	0	0	0	0
Textularia aglutinans	0	0	3	0	0	0	3	1,9	0	0	0	0	0	0	0	0
Triloculina sp.	0	0	0	0	0	0	0	0	0	0	0	0	0	0	0	0
Trochammina inflata	0	12	31	0	0	2	45	**28,7**	9	3	0	0	0	0	12	**27,9**
Trochammina squamata	0	4	19	0	0	0	23	**14,7**	0	0	0	0	0	0	0	0
Total	4	35	78	35	3	2	157	100	33	6	2	0	1	1	43	100
No. of species	3	3	7	3	1	1	-	-	5	3	2	0	1	1	-	-
Wealth (Margalef)	1,4	0,6	1,4	0,6	0	0	-	-	1,1	1,1	1,4	****	****	****	-	-
Evenness (Pielou)	0,9	0,9	0,7	0,9	****	****	-	-	0,9	0,9	1	****	****	****	-	-
Diversity (S-Wiener)	0,6	0,8	1,3	0,9	****	****	-	-	1,2	0,7	0,3	****	0	0	-	-

Caption: MPI1 = Pardo river mangrove from the first campaign at point 1; N = Total; RA = Relative Abundance; Diversity (S - Wiener) = Diversity (Shannon-Wiener).

1.1.9 Taphonomy of the foreheads

In both sampling periods, most of the testes were white or colorless (62.42% in the first and 67.44% in the second) or brown (14.65% and 11.62%). In terms of wear, 46.50% and 65.11% of the foreheads were normal in both samples, although 35.67% and 28.00% showed abrasion (Figure 4.7). As was the case in the estuarine channel, the dominance of white foreskins results from the rapid addition of new foreskins to the sediment (MACHADO et al., 1989), which is confirmed by the predominance of normal foreskins (Figure 4.7).

During the first campaign, there was a high percentage of scorched foreskins, a condition indicative of high-energy environments. However, the presence of the species *A. beccarii*, which is typical of estuarine environments, is evidence of a possible transport of foreskins into the mangrove, justifying the high levels of scorched foreskins in this region (GÓMES; BERNAL, 2013).

Figure 4.7 - Percentages of the types of coloration and wear of foraminifer species in the mangrove zone of the Pardo River, referring to the Nov/2011 and Apr/2012 campaigns.

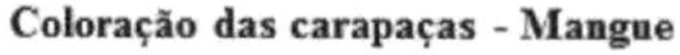

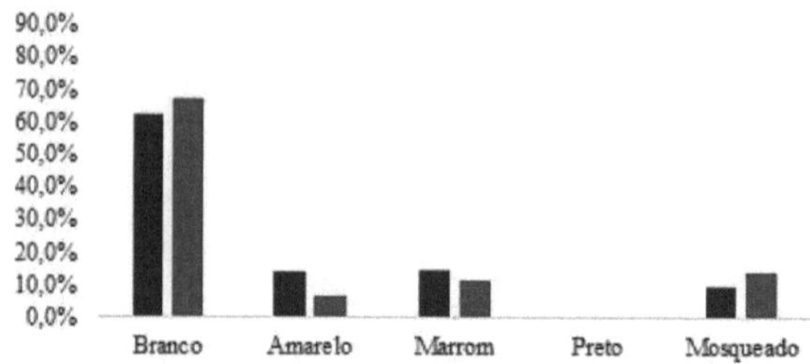

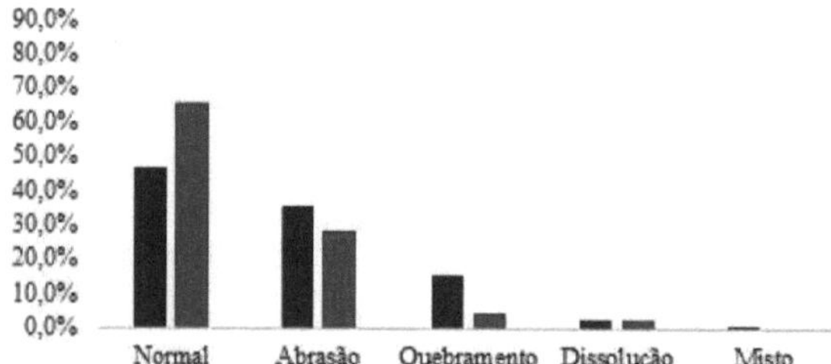

Legenda:
■ Periodo Chuvoso
■ Periodo Seco

4.3.10 Distribution of foraminiferal foreheads

The results of the correlation analysis in the first campaign indicate that there was a positive correlation with the silt fraction, while in the second campaign there was a significant positive correlation between the components fine sand, very fine sand, salinity, Eh, total nitrogen and assimilable phosphorus. These relationships suggest that the granulometric characteristics of the region (smaller granulometric fractions) have a higher organic content and due to the small size of the grains (ALCÂNTARA et al., 2011), consequently favor the associations between the species of living foraminifera and the texture of the sediment (Figure 8).

The dead correlated positively in the first campaign with assimilable phosphorus and the trace metals Mn and Pb. In the second campaign, dead individuals showed a positive correlation with the silt and clay fractions (Figure 8). This is because after death, the organisms tend to settle in environments with lower hydrodynamic energy, in which the fine sand, silt and clay fractions predominate, which in turn, due to their textural characteristics (fine granulometry) have a higher organic content and favor the adsorption of trace metals in the sediment (ONOFRE et al., 2007). In addition, it is possible that the contamination by trace metals in the second campaign led to a reduction in foraminiferal assemblages, compromising the survival of these individuals.

Figure 4.8 - Graphical representation of the Principal Component Analysis, based on biotic, abiotic, metal and nutrient parameters of the sediment in the mangrove zone of the Pardo River, concerning the campaigns carried out in Nov/2011 and Apr/2012

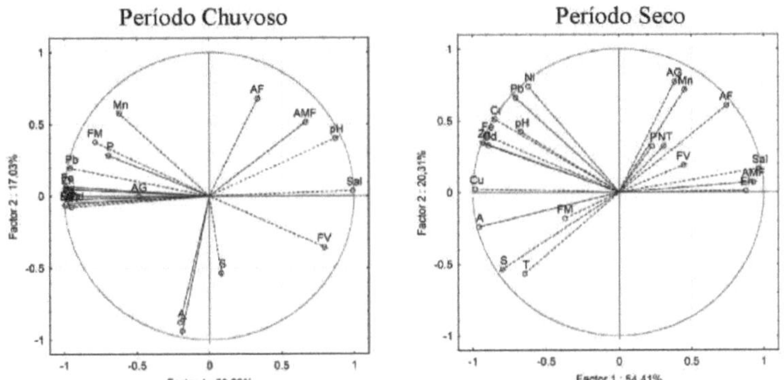

Note: Metals have been symbolized according to the periodic table, but FV = live foraminifera; FM = dead foraminifera; T = temperature; OD = dissolved oxygen; pH = hydrogen potential; Eh = oxidation-reduction potential; AG = coarse sand; AF = fine sand; AMF = very fine sand; S = silt; A = clay; NT = total nitrogen; MO = organic matter.

4.4 CONCLUSIONS

The estuarine channel of the Pardo River is made up of brackish (0.1 to 16.9 - first campaign) and saline (31.0 to 1.7 - second campaign) well-oxygenated waters, but the salinity and temperature values were lower in November 2011 than in April 2012 due to an anomaly in the region's rainfall regime, so that, contrary to expectations, the first campaign was considered to have been carried out in the "rainy season" and the second in the "dry season".

The pH of the sediments in the study region ranged from acidic to neutral and was predominantly reductive during the rainy season and oxidizing during the dry season. Fine sand was the predominant granulometric fraction, confirming the expectation of a low hydrodynamic energy condition in the channel.

The concentrations of assimilable phosphorus, organic matter, organic carbon, nickel, chromium and iron in the canal may come from urban effluents and drainage from agricultural areas upstream, especially during the rainy season, while nitrogen may be associated with urban and industrial effluents. On the other hand, while cadmium concentrations probably come from marine material, since the points with the highest levels were located at the mouth of the canal.

In the Pardo river channel, 467 testes were obtained during the rainy season, generally represented by the species *Ammonia beccarii* (30.6%), *Elphidium excavatum (9.4%), E. poeyanum (5.1%), Haploprhagmoides wilberti (12.2%), Trochammina inflata (19.5%)* and *T. squamata (12.6%). squamata* (12.6%); during the dry season, 96 individuals were obtained, mainly of *A. beccarii*

(7.1%); *A. tepida* (32.1%); *E. excavatum (7.1%)*; *H. wilberti* (10.7%); *T. inflata* (17.9%) and *T. squamata* (7.1%). The presence of the genera *Ammonia* and *Elphidium* at the point (rainy season) indicates a gradual transition from marine conditions, with a high salinity content, possibly due to the period of greater communication with the sea, while the occurrence of the species *H. wilberti, T. squamata* and *T. inflata* indicates estuarine environments of low hydrodynamic energy and a tendency towards hyposalinity at the other points along the channel. In the dry season (second campaign), the reduction and significant absence of foraminiferal tests is possibly due to the high concentrations of cadmium, with the species *Elphidium excavatum* standing out as the most tolerant to trace metal contamination, followed by *A. tepida* and *A. beccarii,* in that order.

The predominance of white foreheads is indicative of a very rapid deposition rate, with a lot of new material being added to the sediment, especially during the dry season, which is corroborated by the dominance of normal foreheads.

In the first campaign, the distribution of live individuals was positively correlated with dissolved oxygen, suggesting that these organisms prefer well-oxygenated environments. For the second campaign, the live individuals showed a positive correlation with salinity, probably indicating that the increase in salinity favored the survival of the species.

The distribution of dead individuals showed a positive correlation in the first campaign with the elements (very fine sand>silt>phosphorus) and in the second campaign with fine sand, given that, after the death of these organisms, the remains are usually deposited in environments with lower hydrodynamic energy, where, consequently, fine-grained sediments predominate which, in turn, due to their small size, favor the deposition of phosphorus.

For the mangrove zone of the Pardo River estuarine channel, salinity and temperature values were not measured during the first campaign due to problems with the equipment, but according to the values measured during the dry period, high salinities (20,0ups to 35.0 ups) and temperatures (25.3°C to 26.6°C) due to an anomaly in the rainfall regime of the region, so that, contrary to expectations, the first campaign was considered to have been carried out during the "rainy season" and the second during the "dry season".

The sediment in the region is slightly acidic to neutral and predominantly oxidizing in both sampling periods. The very fine sand and silt fractions were the predominant granulometric compositions, confirming a depositional environment with lower hydrodynamic energy.

The concentrations of assimilable phosphorus are possibly the result of domestic effluents and shrimp farming activities near the upstream end of the river, as are the concentrations of cadmium, chromium, lead, iron and zinc, which must be originating from agricultural drainage, especially during the rainy season, while nickel may be the result of mining activities and industrial discharges in the region. While nitrogen results from primary biological production in the aquatic system. In the mangrove area of the Pardo river channel, 157 foraminifera were obtained during the rainy season, represented by *Haplophragmoides wilberti* (34.4%), *Quinqueloculina fusca (10.2%), Trochammina inflata* (28.7%) and *T. squamata (14.7%). squamata* (14.7%); in the dry period, 43 foraminifera were obtained, mainly *Ammonia beccarii* (7.0%), *Elphidium excavatum* (23.3%), *H. wilberti* (30.2%) and *T. inflata (*27.9%), and the occurrence of these species indicates an environment with predominantly fine sedimentation and high organic content. While the significant reductions in foraminifera in the second campaign are possibly due to the high concentrations of trace metals, with the species *A. beccarii* and *E. excavatum* standing out as the most tolerant to trace metal levels in the sediments. The predominance of white foreheads is indicative of a very rapid deposition rate, with a lot of new material being added to the sediment, especially during the dry season, which is corroborated by the dominance of normal foreheads.

In both campaigns, the results of the correlation analysis indicate that the distribution of living individuals is positively correlated with the contents of the smallest granulometric fractions (fine sand, very fine sand and silt) and total nitrogen and assimilable phosphorus for the second campaign, showing that sites with fine sedimentation have a higher organic content and consequently a direct association between sediment texture and benthic foraminifera. The distribution of dead individuals showed a positive correlation with the elements (clay>silt>phosphorus) since, after the death of these organisms, the remains are usually deposited in environments with lower hydrodynamic energy, where, consequently, fine-grained sediments predominate, which in turn favor the deposition of phosphorus due to the small size of their grains.

4.5 REFERENCES

AZEVEDO, J.S.; BRAGA, E.S.; FAVARO, D.T.; PERRETTI, A.R.; REZENDE, C.E.; SOUZA, C.M.M. Total mercury in sediments and in Brazilian Ariidae catfish from two estuaries under different anthropogenic influence. **Marine Pollution Bulletin**, v.62, n.12, p.2724-2731, dec 2011.

AGUIAR, P.C.B.; MOREAU, A.M.S.S.; FONTES, E.O. Impacts on the environmental dynamics of the municipality of Canavieiras (BA) with the Resex as a factor of influence. **Revista de Geografia, Meio Ambiente e Ensino,** v.2, n.1, p.61-78, 1°Sem 2011.

BRUNO, R.L.M. Paleoenvironmental reconstruction of the Maricà Lagoon, RJ, based on benthic foraminifera. **Pesquisas em Geociências,** v.40, n.3, p.259-273, sep./dec. 2013.

BALDOTTO, M.A.; CANELLAS, L.P.; ROSA, R.C.C.; RANGEL, T.P.; SALOMÂO, M.S.M.B.; REZENDE, C.E. Oxidizing capacity as an index of sediment organic matter stability according to the fluvial-estuarine gradient of the Paraiba do Sul River. **Quimica Nova,** v.34, n.6, p.973-978, apr. 2011.

BARBOSA, V.P. Benthic foraminifera as biostratigraphic indicators in the upper Quaternary of the Campos Basin. **Revista Brasileira de Paleontologia,** v.13, n.2, p.129-142, May/Aug. 2010.

BERRÊDO, J.F.; DA COSTA, M.L.; VILHENA, M.P.S.P.; SANTOS, J.T. Mineralogy and geochemistry of mangrove sediments from the Amazon coast: the example of the Marapanim river estuary (Parâ). **Revista Brasileira de Geociências**, v.38, n.1, p.24-35, mar. 2008.

BRAZIL. Ministry of the Environment. National Environment Council, CONAMA. **CONAMA Resolution No. 357**, March 17, 2005. Available at:

<http://www.mma.gov.br/port/conama/legiabre.cfm?codlegi=459>. Accessed on: 13 Mar. 2014.

BRAZIL. Ministry of the Environment. National Environment Council, CONAMA. **CONAMA Resolution No. 454**, November 1, 2012. Available at:

<http://www.mma.gov.br/port/conama/legiabre.cfm?codlegi=693> Accessed on: 30 Jan. 2015.

CELINO, J.J. et al. Geochemistry of surface water in the lower reaches of the Una, Pardo and Jequitinhonha rivers, southern Bahia. In: CELINO, J.J.; HADLICH, G.M.; QUEIROZ, A.F.S.; OLIVEIRA, O.M.C. (Orgs.). **Evaluation of coastal environments in the southern region of Bahia:** geochemistry, oil and society. Salvador: EDUFBA, 2014. p.63-76.

CRUZ, F.C. **Trace elements in mangrove substrate from the municipalities of Una, Canavieiras and Belmonte, Bahia**. 2012. 110 f. Dissertation (Master's Degree in Geochemistry) - Institute of Geosciences, Federal University of Bahia, Salvador, 2012.

CARVALHO, P.V.; SANTOS, P.J.; BOTTER-CARVALHO, M.L. Assessing the severity of disturbance for intertidal and subtidal macrobenthos: the phylum-level meta-analysis approach in tropical estuarine sites of northeastern Brazil. **Marine Pollution Bulletin,** v.60, n.6, p.873887, jun. 2010.

CLARKE, K.R.; WARWICK, R.M. **Change in marine communities:** an approach to statistical analysis and interpretation. 2nd ed: Plymouth Marine Laboratory, 2001. 172p.

CARR, M.R. **(Plymouth Routines in multivariate Ecological Research) User Manual**. 2nd ed. [s.l.]: Plymouth Marine Laboratory, 1996. 169p.

DUQUET, M. **Ciências da vida:** glossàrio de ecologia fundamental. Porto: Porto Editora, 2007. 128p.

DONNICI, S.; SERANDREI-BARBERO, R.; BONARDI, M.; SPERLE, M. Benthic foraminifera as proxies of pollution: The case of Guanabara Bay (Brazil). **Marine Pollution Bulletin**, v.64, n.10, p.2015-2028, oct. 2012.

DAUVIN, J.C. Paradox of estuarine quality: Benthic indicators and indices, consensus or debate for the future. **Marine Pullution Bulletin,** v.55, n. 1-6, p.271-281, 2007.

DAJOZ, R. **General ecology**. 2.ed. Petrópolis: Vozes, 1983. 472p.

DELGADO, A.F. **Gestión de zonas costeras con técnicas estocàsticas multicriterio.** 2012. 197f. Thesis (Doctorate in Engineering) - Escuela Técnica Superior de Ingenieros de Caminos, Canales y Puertos Universidad de Granada, 2012.

EICHLER, P.P.B.; EICHLER, B.B.; Gupta, B.S.; Rodrigues, A.R. Foraminifera as indicators of marine pollutant contamination on inner continental shelf of southern Brazil. **Marine Pollution Bulletin,** v.64, n.1, p.22-30, jan. 2012.

ESCOBAR, N.F.C.; CELINO, J.J.; NASCIMENTO, R.A. Metals in surface water, suspended particulate matter and bottom sediment in the lower reaches of the Una, Pardo and Jequitinhonha rivers. In: CELINO, J.J.; HADLICH, G.M.; QUEIROZ, A.F.S.; OLIVEIRA, O.M.C. (Orgs.). **Evaluation of coastal environments in the southern region of Bahia:** geochemistry, oil and society. Salvador: EDUFBA, 2014. 77-98p.

ESCOBAR, N.F.C. **Geochemistry of surface water and bottom sediment in the lower reaches of the Una, Pardo and Jequitinhonha rivers, Southern Bahia, Brazil.** 2013. 121f. Dissertation (Master's Degree in Geochemistry) - Institute of Geosciences, Federal University of Bahia, Salvador, 2013.

EICHLER, P.P.B.; EICHLER, B.B.; MIRANDA, L.B.; RODRIGUES, A.R. Modern foraminiferal facies in a subtropical estuarine channel, Bertioga, Sao Paulo, Brazil. **Journal of Foraminiferal Research,** v.37, n.3, p.234-247, jul. 2007.

FÉLIX, A.A.; BAQUERIZO, J.M.; SANTIAGO, M.A. Losada. Coastal zone management with stochastic multi-criteria analysis. **Journal of Environmental Management,** v.112, p.252-266, dec. 2012.

FAGNANI, E.; GUIMARAES, J.R.; MOZETO, A.A.; FADINI, P.S. Acid volatile sulfides and simultaneously extracted metals in the assessment of freshwater sediments. **Quimica Nova,** v.34, n.9, p. 1618-1628, may. 2011.

FRONTALINI, F.; COCCIONI, R. Benthic foraminiferal for heavy metal pollution monitoring: A case study from the central Adriatic Sea coast of Italy. **Estuarine Coastal and Shelf Science,** v.76, n.2, p.404-417, jan. 2008.

GÓMES, E.; BERNAL, G. Influence of the environmental characteristics of mangrove forests on recent benthic foraminifera in the Gulf of Urabà, Colombian Caribbean. **Marine Sciences,** v.39, n.1, 2013.

GOMES, R.C.T. **Caracterizaçao da fauna de foraminiferos da zona euhalina do Estuàrio do Rio Jacuipe - Camaçari-Ba**. 2010. 168f. Dissertation (Master's in Geology) - Institute of Geosciences, Federal University of Bahia, Salvador, 2010.

HORTELLANI, M.A.; SARKIS, J.E.S.; ABESSA, D.M.S.; SOUSA, E.C.P.M. Assessment of metallic element contamination in sediments from the Santos - Sao Vicente Estuarine System. **Quimica Nova,** v.31, n.1, p.10-19, 2008.

IBGE - Brazilian Institute of Geography and Statistics. **Population Count.** 2014. Available at: <http://www.cidades.ibge.gov.br>. Accessed on: March 13, 2014.

KNOPPERS B.; MEDEIROS P.R.P.; SOUZA W.F.L.; JENNERJAHN T. The Sao Francisco Estuary, Brazil. **Hdb Env Chem.,** v.5, p.51-70, 2006.

MESSIAS, I.A.M.; COUTO, E.G.; AMORIM, R.S.S.; JOHNSON, M.S.; JUNIOR, O.B.P. Continuous monitoring of redox potential and complementary variables in a hyperseasonal environment in the Northern Pantanal. **Revista Brasileira de Ciências do Solo,** v.37, p.632-639, 2013.

MACHADO, A.!; ARAUJO, T.M.F.; De ARAÙJO, H.A.B.; FIGUEIREDO, S.M.C. Bathymetric and taphonomic analysis of the foraminiferal microfauna of the continental shelf and slope of the Municipality of Conde, Bahia. **Cadernos de Geociências,** v.9, n.2, p. 157-172, nov. 2012.

MARTINS, R.V.; FILHO, P.F.J.; ESCHRIQUE, S.A.; LACERDA, L.D. Anthropogenic sources and distribution of phosphorus in sediments from the Jaguaribe River estuary, NE, Brazil. **Brazilian Journal of Biology,** v.71, n.3, p.673-678, aug. 2011.

MURRAY J. **Ecology and Applications of Benthic Foraminifera.** Cambridge University Press, 2006. 426p.

ONOFRE, C.R.E.; CELINO, J.J.; NANO, R.M.W.; QUEIROZ, A.F.S. Bioavailability of trace metals in mangrove sediments from the northern portion of Todos os Santos Bay, Bahia, Brazil. **Revista de Biologia e Ciências da Terra,** v.7, n.2, p. 65-82, 2.Sem. 2007.

PERH-BA. State Water Resources Plan for the State of Bahia. **Diagnosis and Regionalization.** Salvador, 2003 (Final report of stage I). Available at: <http://biblioteca.inga.ba.gov.br. Accessed on>. Accessed on: March 14, 2014.

PADIAL, P.R. **Quality, spatial heterogeneity and bioavailability of metals in the sediment of a**

tropical urban eutrophied reservoir (Guarapiranga Reservoir, SP). 2008. 109 f. Dissertation (Master of Science) - Institute of Biosciences, University of Sao Paulo, 2008.

QUEIROZ, A.F.S.; OLIVEIRA, O.M.C. **Geoenvironmental diagnosis of mangrove areas and development of technological processes applicable to the remediation of these areas:** subsidies for an impact prevention program in areas with potential for oil activities in the southern coastal region of the State of Bahia (PETROTECMANGUE-BASUL). Salvador: EDUFBA, 2013. 148p. (Technical report).

ROLLS. R.J.; LEIGH. C.; SHELDON. F. Mechanistic effects of low-flow hydrology on riverine ecosystems: ecological principles and consequences of alteration. **Freshwater Science,** v.31, n.4, p.1163-1186, dec. 2012.

REEF, R.; FELLER, I.C.; LOVELOCK, C.E. Nutrition of mangroves. **Tree Physiology,** v.30, p.1148-1160, apr. 2010.

RODRIGUES, A.R.; MALUF, J.C.C.; BRAGA, E.S.; EICHLER, B.B. Recent benthic foraminiferal distribution and related environmental factors in Ezcurra Inlet (King George Island, Antarctica). **Antarctic Science,** v.22, n.4, p.343-360, 2010.

RAMOS, M.G.M.; GERALDO, L.P. Assessment of avicennia schaueriana, laguncularia racemosa and rhizophora mangle plant species as bioindicator of heavy metal pollution in mangrove environments. **Engenharia Sanitària e Ambiental,** v. 12, n.4, p.440-445, dec. 2007.

STIEGLITZ. T.C.; CLARK. J.F.; HANCOCK. G.J. The mangrove pump: The tidal flushing of animal burrows in a tropical mangrove forest determined from radionuclide budgets. **Geochimica et Cosmochimica Acta.,** v.102, p.12-22, mar. 2013.

SATYANARAYANA. B.; MULDER. S.; JAYATISSA. L.P.; DAHDOUH-GUEBAS. F. Are the mangroves in the Galle-Unawatuna area (Sri Lanka) at risk? A social-ecological approach involving local stakeholders for a better conservation policy. **Ocean and Coastal Management,** v.71, p.225-237, jan. 2013.

SANTOS. E.S.; JENNERJAHN. T.; LEIPE. T.; MEDEIROS. P.R.P.; SOUZA. W.F.L.; KNOPPERS. B.A. Origin of sedimentary organic matter in the delta-estuary of the Rio Sao Francisco, AL/SE - Brazil. **Geochimica Brasiliensis,** v.27, n.1, p.37-48, 2013.

SAMPAIO, N.; VARGAS, M.A.M. The landscape of the Pardo river unveiled by riverine community in southwestern Bahia: Talks between the perceived and lived. **Revista Eletrônica Ateliê Geogràfico,** v.4, n.4, p.147-177, dec. 2010.

SANTOS, P.S., MARQUES, A.C.; ARAUJO, A. Remnants of coastal vegetation in the southeast region of Bahia - municipalities of Una and Canavieiras. In: MOSTRA DE TALENTO CIENTiFICO, 1., 2002, Curitiba. **Proceedings...** Curitiba: GIS Brasil, 2002. p.1-6.

TEODORO, A.C.; DULEBA, W.; GUBITOSO, S. Multidisciplinary study (geochemistry and foraminiferal associations) to characterize and evaluate anthropogenic interventions in Araçâ Bay. **Geosciences,** v.11, n.1, p.113-136, apr. 2011.

TEODORO, A.C.; DULEBA, W.; GUBITOSO, S.; PRADA, S.M.; LAMPARELLI, C.C.; BEVILACQUA, J.E. Analysis of foraminifera assemblages and sediment geochemical properties to characterize the environment near Araçâ and Saco da Capela domestic sewage submarine outfalls of Sao Sebastiao Channel, Sao Paulo State, Brazil. **Marine Pollution Bulletin,** v.60, n.4, p.536-553, dec. 2010.

TEODORO, A.C.; DULEBA, W.; LAMPARELLI, C.C. Foraminiferal associations and textural composition of the region near the Cigarras domestic sewage submarine outfall, Sao Sebastiao Channel, SP, Brazil. **Pesquisas em Geociências,** v.36, n.1, p.467-475, jan./abr. 2009.

VIDOTTI, E.C.; ROLLEMBERG, M.C.E. Algae: from aquatic environment economy to bioremediation and analytical chemistry. **Quimica Nova,** v.27, n.1, p.139-145, 2004.

XAVIER, A. L. S. **Paleotectonics of the provenance areas and petrography of the Salobro Formation, Rio Pardo Basin - Bahia.** 2009. 89f. Monograph (Graduation in Geology) - Institute of Geosciences, Federal University of Bahia, Salvador, 2009.

YAKUTINA, O.P. Phosphorus content in sediment and eroded soils in the southeastern part of West Siberia. **Agriculture, Ecosystems and Environment,** v.140, n.1-2, p.57-61, jan. 2011.

ZOURARAH, B.; MAANAM, M.; ROBIN, M. Source Contributions to Heavy Metal Fluxes into the Loukous Estuary (Moroccan Atlantic Coast). **Journal of Coastal Research,** v.28, n.1, p.174-183, 2012.

5 STUDY OF GEOCHEMICAL CHARACTERIZATION AND ASSOCIATIONS OF PARALLIC FORAMINIPHEROS IN THE CHANNEL AND MANGUEZAL ZONE OF THE JEQUITINHONHA RIVER ESTUARY, SOUTHERN LITTORAL OF BAHIA

Summary

Paralytic foraminiferal associations, besides being abundant and widely distributed geographically, take advantage of the intrinsic peculiarities of these organisms in storing in their foreheads characteristics of the places where they have lived. In this context, the testes of these organisms were related to the levels of trace metals in the sediment of the mangrove zone of the estuarine channel of the Jequitinhonha River, on the southern coast of Bahia, with the aim of assessing whether the levels of these elements are affecting the microfauna. Two seasonal samplings were carried out (November/2011, April/2012) to collect 10 samples of bottom sediment in the channel and 6 samples of surface sediment near the *Avicennia* specimens in the mangrove, totaling 32 samples. A total of 281 foraminiferal specimens were obtained from the canal region in the first campaign (4.98% of the specimens were collected alive, and none of the specimens were malformed) belonging to 10 species, of which *Trochammina inflata*, *Haplophragmoides wilberti* and *Ammonia beccarii* stand out as the main species. In the second sampling, no foraminifera were recorded due to the low salinities, making it impossible for even the most resistant species to survive. In the mangrove zone, in the first campaign, 116 samples (5.1% live; 0.0% anomalous) of 10 species were obtained, with *Haplophragmoides wilberti*, *Ammonia beccarii*, *Quinqueloculina fusca* and *Q. venusta standing* out. In the second campaign, only 143 testes were recorded (0.70% live; 0.0% anomalies) from 10 species, with only 4 species standing out, namely: *H. wilberti*; *Trochammina inflata* and *Q. fusca, and* the increase in the number of testes was due to the increase in salinity. In addition, it was only observed in the first campaign in the Pardo river channel that only the metal lead (points 2, 3, 4, 5 and 6) showed values above the limits established by the reference bodies (CONAMA - Brasil, 2012) and the Canadian Council of Ministers of the Environment Canadian Environmental Quality Guidelines (CCME, 1998). In view of this, it was observed that the high concentrations of lead in the rainy season are possibly related to the drainage of fertilizers and chemical compounds used in agricultural development. In the Mangrove Zone, high concentrations of the metals copper (point 2), lead (point 1) and cadmium (point 2) were observed only in the dry season, possibly due to the reduction in flow, increasing concentrations in the water and consequently enabling their deposition in the sediments. Thus, the presence of the species *A. beccarii* and *H. wilbert is* probably due to the fact that they are opportunistic species, since both are commonly found in places with nutrient inputs. While the species *T. inflata* is often found in mangrove areas, due to the predominance of sediments with a smaller grain size and higher nutrient content.

Keywords: Foraminifera, Belmonte, Estuaries, Brazil

5.1 INTRODUCTION

Estuaries are characterized as coastal environments located at the land-sea interface, where they act as receptacles for substances and products resulting from human interactions, which can cause degradation of waters and ecosystems (TEODORO et al., 2010). Consequently, the direct or indirect introduction of these interferences by man can lead to various levels of contamination of estuarine systems, with serious effects on natural resources, risks to human health, barriers to fishing activities, deterioration of water quality and reduction of its natural beauty (VIANA et al., 2012).

In parallel to this is the mangrove swamp, defined as a tropical ecosystem with dominant intertidal vegetation, which contributes significantly to the survival of various species of animals and plants, as well as participating in the primary production of global biomass (BOUILLON et al., 2008). However, as population growth has increasingly shifted towards coastal areas, domestic sewage has become one of the main causes of marine pollution (LAMPARELLI, 2007).

Thus, environmental studies using foraminifera as bioindicators have become increasingly common, as these organisms are highly sensitive to variations in physical and chemical factors in the environment, have short life cycles and high biodiversity (COSENTINO et al., 2013). In addition, the responses of these protists can include extinctions of some species, changes in agglutination with increasing density and low diversity (FRONTALINI; COCCIONI, 2008).

The aim of this study was to describe the prevailing conditions in the estuarine channel and mangrove zone of the Jequitinhonha River, on the southern coast of the state of Bahia, using the characteristics of paralic foraminifera and geochemical data from the sediment.

5.2 MATERIALS AND METHODS

The study area corresponds to the channel and mangrove area of the Jequitinhonha River estuary, on whose banks lies the municipality of Belmonte, located on the Dendê Coast, on the southern coast of the state of Bahia (Figure 5.1). The Jequitinhonha river basin has a population of 23,471 inhabitants, covering an area of 1,970.142 km^2 , which includes part of the northeast of the state of Minas Gerais (66,319 km2) and a small sector of the southeast of Bahia (3,996 km2) (IBGE, 1997). It is bordered to the north by the Pardo river basin, to the west by the São Francisco river basin, to

D. Drying at 60 C°   E. Flotation of the foreheads of   F. Ideutification of foraminifera.
                                      Ioiamiiiiferos.

In the Foraminifera Study Group Laboratory (LGEF), using a stereomicroscope, the specimens were removed from the filter paper and fixed with organic glue on microfossil slides. The foraminifera were identified on the basis of specialized literature and, during this procedure, information regarding their colouring and state of preservation was recorded (MACHADO et al., 2012) (Figure 5.2F). The occurrence of forehead anomalies was also recorded, when present.

5.2.3 Sediment analysis

The results of the analyses of the sediment from the channel and the mangrove swamp of the Jequitinhonha River were provided by NEA researchers, so the procedures for carrying out the analyses of granulometry, assimilable phosphorus, organic matter, organic carbon, total nitrogen and metals are described in Celino et al. (2014), Escobar et al. (2014) and Cruz (2012).

5.2.4 Statistical analysis

For the descriptive analysis of the foraminifer fauna, relative abundance was calculated (AB'SABER et al., 1997) and the following classes were adopted: main (abundances >5%), accessory (4.9-1%) and trace (<1); and constant (occurrences >50%), accessory (49-25%) and accidental (<24%) (DAJOZ, 1983). In addition, the richness (Margalef index), evenness (Pielou index) and diversity (Shannon-Wiener index) indices were calculated using the Primer 6.0 program (CARR, 1996) (CLARKE; Warwick, 2001).

To establish the correlation between foraminifer distribution and environmental variables, the faunal results were combined with those of the physical-chemical parameters and the granulometric and geochemical analyses to create a data matrix, from which correlation and principal component analyses were carried out using the Statistica 8.0 program (StatSoft, 2007).

5.3 RESULTS AND DISCUSSIONS

The results of the physico-chemical parameters monitored in this study showed a rainfall anomaly in the months of the sampling, so that the rainfall was higher (332mm per month) during the first campaign (Nov 2011), which was characterized as a rainy season, while in the second sampling (Apr 2012) the rainfall rate was reduced (50mm monthly), characterizing it as a dry season (ESCOBAR, 2013), which certainly influenced the values of salinity, temperature and other parameters recorded during the collections. Therefore, for the purposes of this study only, the first campaign will be considered to have been carried out during the "rainy season" and the second during the "dry season".

5.3.1 Jequitinhonha River estuarine channel: physico-chemical parameters

The results of the physico-chemical parameters monitored were in accordance with the limits established in CONAMA Resolution 357 of March 17, 2005 (BRASIL, 2005), and based on the salinity data, it is possible to observe salinity values of less than 0.5‰ between points 2 and 4 (first campaign) and at points 1 and 3 to 6 (second campaign),5‰ between points 2 and 4 (first campaign) and points 1 and 3 to 6 (second campaign), while salinity values of more than 0.5‰ and less than 30.0‰ were only recorded at point 1 (rainy season) and point 2 (dry season) (Figure 5.3). Contrary to expectations, salinity was absent at the other points analyzed, and salinity and temperature values were lower in the first (0.0 and 0.7 ups, and 25.7 to 27.0 °C in Nov/2011) than in the second (0.0 and 8.6 ups, and 27.4 to 30.1 °C in Apr/2012) sampling campaign (Figure 5.3).

Thus, during the sampling months, there was an anomaly in the local rainfall, which showed a high rainfall index (332mm per month) during the first campaign, behaving like a rainy period, while in the second sampling the rainfall rate was reduced (50mm per month), characterizing it as a dry season, which certainly influenced the values of the other physicochemical parameters, especially temperature and salinity recorded during the sampling (ESCOBAR, 2014). Therefore, for the purposes of this research only, the first campaign will be considered the "rainy season" and the second the "dry season".

In both campaigns, the estuarine waters of the Jequitinhonha River channel were slightly oxygenated (13.5 to 18.3 mg/L^{-1} in the first campaign and 4.6 to 9.0 mg/L^{-1} in the second - Figure 5.3), and the lower values recorded in the dry season may be related to the reduction in water volume and increase in temperature (AZEVEDO et al., 2011) in the month of sampling.

The pH values of the sediment were slightly acidic to neutral (6.5 to 8.1 in the first campaign and between 6.7 and 7.9 in the second - Figure 5.3). In the first campaign, when we analyze the location of the collection points in the estuary, we see slightly more acidic conditions from point 7 onwards, confirming the expectation that the points near the mouth are more influenced by the marine system than the point upstream, which is under strong fluvial influence (VEIGA, 2010), a condition that is repeated in the second campaign.

For the Eh values of the sediment, in the first campaign negative values were obtained up to point 9, configuring it as a reducing environment, while point 10 was under an oxidizing condition. In the second campaign, all the points had negative Eh values (Figure 5.3).

5.3.2 Sediment analysis

In both samples, fine sand was the predominant particle size, confirming the expectation of a condition of lower hydrodynamic energy in the channel (Figure 5.3).

In the dry period (April 2012 sampling), there was a reduction in assimilable phosphorus values (2.5mg/L to 168.8mg/L in the first campaign and 1,2mg/L to 30.2mg/L in the second), organic matter (0.3mg/L to 2.5mg/L and 0.2mg/L to 0.8mg/L) and organic carbon (0.2mg/L to 1.4mg/L and 0.1 to 0.4mg/L). Only the total nitrogen values (0.1mg/L to 1.1mg/L and 0.1mg/L to 2.6mg/L) increased (Table 3.1).

Figure 5.3 - Values of the physicochemical parameters, salinity, temperature, dissolved oxygen (D.O.) of the water and hydrogen potential (pH), oxidation potential (Eh) and granulometric fractions of the sediment of the estuarine channel of the Jequitinhonha River, relative to the campaigns of Nov/2011 (rainy season) and Apr/2012 (dry season).

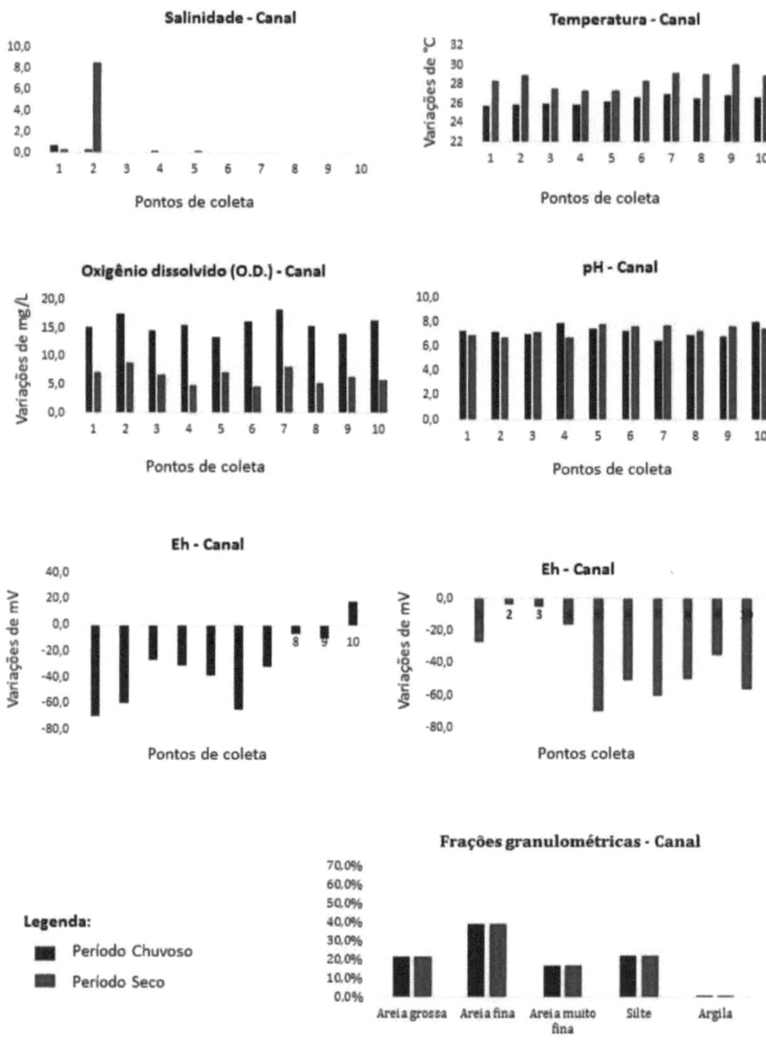

These results show that the concentrations of assimilable phosphorus, organic matter and organic carbon may be the result of agricultural drainage from nearby upstream regions, which becomes even more intense during the rainy season, while nitrogen probably comes from urban and industrial effluents, which are more concentrated during the dry season (CARVALHO et al., 2010).

For comparisons and analysis of the consequences of the high levels of trace metals in the sediments in the study area, we opted to use the limits established by CONAMA Resolution No. 454/2012 (BRASIL, 2012) and the *Canadian Environmental Quality Guidelines* (CCME, 1998), which showed that the concentration of Ni, Cr, Cu, Pb and Cd (first campaign) in the sediment is low and therefore should not be causing adverse effects to the biota of the Jequitinhonha river channel (Table 5.2). However, these standards do not set limits for Mn and Fe, which makes it impossible to assess the effect of these elements. However, comparing the results obtained for the concentration of the metal Cd at points 3 and 4 (13.48 mg.Kg^{-1} and 0.46 mg.Kg^{-1}) in the first campaign and all the points (0.45 mg.Kg^{-1} point 3 - highest value) in the second campaign, it was observed that these showed concentration values above the limits of the reference bodies. This information suggests that the point source of cadmium contamination at point 3, close to downstream during the rainy season, comes from marine material (marine phosphate and aquatic systems).

5.3.3 Foraminifera fauna

In the Jequitinhonha river channel, 281 individuals were obtained during the rainy season (Nov/2011), of which 14 were alive. There were no anomalous species, and 5 species were identified: *Trochammina inflata* (37.01%), *Haplophragmoides wilberti* (22.79%) and *Ammonia beccarii* (18.50%) as the main ones (Table 5.3). The presence of *T. inflata* is indicative of environments with low hydrodynamic energy and brackish waters, while *H. wilberti* is preferentially distributed in fine-grained, nutrient-rich substrates. *A. beccarii, on the* other hand, is the most tolerant of major changes in temperature and salinity, as well as having a preference for sediment with nutrient inputs, corroborating the ecological studies of benthic foraminifera carried out by Murray (2006).

In the dry season, the non-occurrence of foraminifera is due to the low salinity values, which made it impossible for even the most resistant species to survive (GÓMES; BERNAL, 2011) (Tables 5.1 and 5.3).

Table 5.1 - Physicochemical, granulometric and nutrient data of the water and sediment of the estuarine channel of the Jequitinhonha River concerning the first (I - Nov/2011) and second (II - Apr/2012) sampling campaigns and limits of CONAMA Resolution 357 of March 17, 2005 (Brazil, 2005).

| Samples | Water parameters | | | Sediment Analysis | | | | | | | | |
	T (°C)	Salt (ups)	OD (mg/L-1)	pH	Eh (mV)	GA (%)	PA (%)	MFA (%)	S (%)	A (%)	NT (mg/L)	P (mg/L)
CJI1	25,7	0,7	15,1	7,3	-70	13,1	0,3	35,2	51	0,5	0	168,8
CJI2	25,8	0,3	17,4	7,2	-60	26,8	0	23,4	48,9	0,9	0	133,7
CJI3	26	0,1	14,4	7,1	-27	13,7	76,7	6,3	3,2	0	1,1	132,2
CJI4	25,8	0,2	15,6	7,9	-31	51,6	42,2	3,6	2,6	0	0,7	14,8
CJI5	26,2	0	13,5	7,4	-39	1,9	0,4	40,6	56,7	0,4	1,1	154,6
CJI6	26,7	0	16,1	7,3	-65	55,9	37,4	4,8	1,9	0	0,1	29,3
CJI7	27	0	18,3	6,5	-32	7,4	80,7	8,2	3,7	0	0,1	22,2

Samples	T	O.D	pH	Eh	AG	AF	AMF	S	A	P	NT
CJI8	26,5	0	15,4	7	-7	21,2	64,1	9,3	5,3	0,1	0	2,5
CJI9	26,9	0	13,9	6,9	-10	37,7	51,7	5,5	5	0	0	0
CJI10	26,7	0	16,2	8	18	22,1	21,3	24,2	32,2	0	0,4	2,5
Average	26,33	0,13	15,59	7,26	35,9	25,14	37,48	16,11	21,05	0,19	0,35	66,06
CJII1	28,4	0,3	7,1	7	-27	26,9	63,7	6	3,4	0	0,3	30,2
CJII2	28,9	8,6	9	6,7	-4	25,8	30,8	23,7	19,7	0,1	0,6	4,9
CJII3	27,6	0,1	6,7	7,2	-5	35,6	56	4,5	3,9	0	2,6	2,4
CJII4	27,4	0,1	4,7	6,7	-16	64,8	27,2	4,7	3,3	0	0,2	4,9
CJII5	27,4	0,2	7,1	7,9	-70	37,1	55,6	4,1	3,3	0	0,1	0
CJII6	28,4	0,1	4,6	7,7	-51	44,9	46,8	5,1	3,2	0	0,6	1,2
CJII7	29,2	0	8,2	7,7	-60	53,2	40,1	3,9	2,8	0	0,1	2,5
CJII8	29,1	0	5,2	7,3	-50	52	41,8	3,9	2,2	0	0,6	2,5
CJII9	30,1	0	6,3	7,6	-35	34,4	55,2	6,6	3,8	0	0,6	2,6
CJII10	29	0	5,8	7,5	-56	34,8	57,4	4,7	3,1	0	0,3	2,5
Average	28,55	0,94	6,47	7,33	37,4	40,95	47,46	6,72	4,87	0,01	0,6	5,37
CONAMA												
SWEET	< 40,0	< 5,0	5	-	-	-	-	-	-	-	-	-
SALINE	-	> 30,0	6	-	-	-	-	-	-	-	-	-
SALOBRA	-	0,5 a 30,0	5	-	-	-	-	-	-	-	-	-

Legend: CJI1 = Jequitinhonha river channel from the first campaign; T = Temperature; O.D = Dissolved Oxygen; pH = Hydrogen Potential; Eh = Oxirreduction Potential; AG = Coarse Sand; AF = Fine Sand; AMF = Very Fine Sand; S = Silt; A = Clay; P = Phosphorus; NT = Total Nitrogen; CONAMA = National Environmental Council.

Table 5.2 - Concentrations of metals (in mg. Kg -1) in the sediment of the estuarine channel of the Jequitinhonha River, concerning the first (I - Nov/2011) and second (II - Apr/2012) sampling campaigns and reference values from CONAMA Resolution No. 454/2012 (Brazil, 2012) and the *Canadian Environmental Quality Guidelines* (CCME, 1998).

Samples	Ni	Mn	Fe	Cr	Cu	Pb	Cd
CJI1	0,01	19,98	694	0,02	0,01	0,8	0
CJI2	0,01	11,37	2313	0,4	0,16	1,11	0,01
CJI3	0,01	330,03	327	0,05	0,01	9,24	**13,48**
CJI4	13,17	198,89	16350	19,6	7,31	10,5	0,46
CJI5	5,51	326,49	20488	22,86	8	13,06	0,04
CJI6	0,01	55,1	935	1,12	0,04	0,83	0
CJI7	4,14	23,27	2632	2,05	0,8	2,06	0
CJI8	0,01	53,14	832	1,34	0,01	0,69	0,01
CJI9	0,01	36,58	969	0,05	0,01	0,32	0
CJI10	7,88	162,52	6885	9,93	1,08	3,71	0,01
Average	3,076	121,737	5242,5	5,742	1,743	4,232	1,401
CJII1	0,32	20,48	420	0,47	0,23	0,01	0,38
CJII2	0,62	26,59	986	1,19	0,37	0,52	0,38
CJII3	2,19	110,1	3628	5,41	1,8	1,86	0,45
CJII4	0,43	14,64	660	0,56	0,32	0,61	0,43
CJII5	0,22	20,46	328	0,21	0,08	0,01	0,38
CJII6	0,06	14,72	407	0,24	0,07	0,9	0,42
CJII7	0,32	15,84	544	0,6	0,14	0,95	0,39
CJII8	0,06	19,36	646	0,54	0,15	0,01	0,43
CJII9	0,07	20,19	344	0,05	0,23	0,25	0,41
CJII10	0,15	23,02	436	0,51	0,06	0,01	0,38
Average	0,44	28,54	839,9	0,97	0,34	0,51	0,4
LD	0,00415	0,00111	0,02586	0,01632	0,00336	1,49	0,00095
CONAMA N1	20,9	n.a.	n.a.	81	34	46,7	1,2
CONAMA N2	51,6	n.a.	n.a.	370	270	218	7,2

Legend: LOD = Limit of Detection; N1 and ISQG= threshold below which there is a lower probability of adverse effects on biota; N2 and PEL= threshold above which there is a higher probability of adverse effects on biota; n.a. = not determined.

In the rainy season, the richness values ranged from 0.5 to 1.1; the evenness values were all above 0.5, suggesting the absence of dominance, as observed in statistical studies of marine communities by Clarke and Warwick (2001). The diversity values ranged from 0.3 to 1.3 (Table 5.3), which are considered low when compared to those obtained by Gomes (2010) in the Jacuipe River estuary (BA) (1.0 to 4.4); by Teodoro (2009 and 2010) in the Sâo Sebastiâo Channel, SP (0.3 to 2.8) and Araça (2.4 to 2.9), also near the Sâo Sebastiâo Channel, SP. In the dry season, the lack of foraminifer records made it impossible to calculate the richness, evenness and diversity indices.

Table 5.3 - Absolute (N) and relative (AR) abundances and ecological indices of foraminifer species recorded in the estuarine channel of the Jequitinhonha River, concerning the first (I - Nov/2011) and second (II - Apr/2012) sampling campaigns

Canaldorio Jequitinhonha	JI1	JI2	JI3	JI4	JI5	JI6	JI7	JI8	JI9	JI10	N	AR	JII1	JII2	JII3	JII4	JII5	JII6	JII7	JII8	JII9	JII10	N	AR
Ammonia beccarii	47	2	0	0	2	1	0	0	0	0	52	18,5	0	0	0	0	0	0	0	0	0	0	0	0
Ammonia parkinsoniana	0	4	0	4	0	3	2	0	0	0	13	4,6	0	0	0	0	0	0	0	0	0	0	0	0
Elphidium sagrum	35	0	0	0	1	0	0	0	0	0	36	12,8	0	0	0	0	0	0	0	0	0	0	0	0
Haplophragmoides wilberti	64	0	0	0	0	0	0	0	0	0	64	22,8	0	0	0	0	0	0	0	0	0	0	0	0
Quinqueloculina fusca	2	0	0	1	9	0	0	0	0	0	12	4,3	0	0	0	0	0	0	0	0	0	0	0	0
Trochammina inflata	103	0	0	0	1	0	0	0	0	0	104	37	0	0	0	0	0	0	0	0	0	0	0	0
N per point	251	6	0	5	13	4	2	0	0	0	281	100	0	0	0	0	0	0	0	0	0	0	0	0
No. of species	5	2	0	2	4	2	2	0	0	0	-	-	0	0	0	0	0	0	0	0	0	0	-	-
Wealth	0,7	0,6	***	0,6	1,2	0,7	0	***	***	***	-	-	0	0	0	0	0	0	0	0	0	0	-	-
Evenness (Pielou)	0,8	0,9	***	0,7	0,7	0,8	***	***	***	***	-	-	0	0	0	0	0	0	0	0	0	0	-	-
Diversity (S-Wiener)	1,3	0,5	***	0,3	0,7	0,3	0	***	***	***	-	-	0	0	0	0	0	0	0	0	0	0	-	-

Legend: JI1 = Jequitinhonha River from the first campaign at point 1; N = Total; RA = Relative Abundance; Diversity (S - Wiener) = Diversity (Shannon-Wiener).

5.3.4 Taphonomy of the foreheads

During the rainy season, most of the testes were white or colorless, 63.70%, or yellow, 25.62%. As for wear, 44.48% of the foreheads were normal, but 29.89% showed signs of abrasion (Figure 5.4).

The predominance of white cores is indicative of a very rapid deposition rate, with a lot of new material being added to the sediment, especially during the dry season, which is corroborated by the dominance of normal cores. However, the fact that abrasion is the most frequent type of wear shows a high-energy environment (MACHADO et al, 2012), contrary to the granulometric data which denote a low-energy environment for the estuarine channel.

Figure 5.4 - Percentages of the types of coloration and wear of foraminifera in the estuarine channel of the Jequitinhonha River, referring to the Nov/2011 and Apr/2012 campaigns

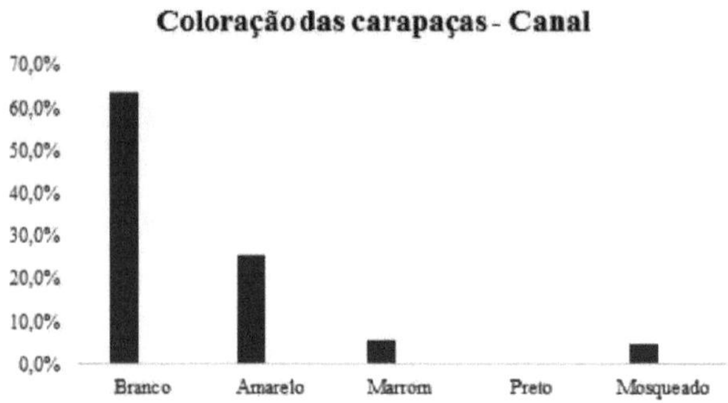

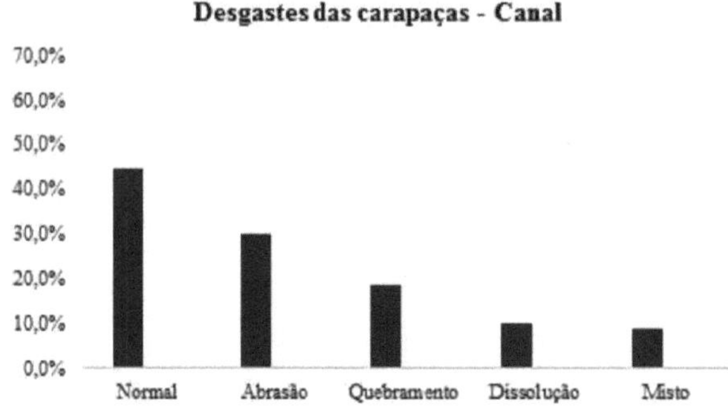

Legenda:
■ Periodo Chuvoso

5.3.5 Distribution of foraminifera

In the first campaign, the distribution of live and dead individuals was positively correlated with the fraction of very fine sand, silt, clay and assimilable phosphorus (Figure 5.5), indicating that fine sediment has a higher organic content, showing a direct association between sediment texture and benthic foraminifera, corroborating the results presented by Machado et al (2012), in taphonomic studies in the municipality of Conde - BA. For the second campaign, the absence of foraminifera, possibly due to the low salinity values and contamination by trace metals, made it impossible to carry out a correlation analysis of the species with the abiotic parameters (Figure 5.5).

84

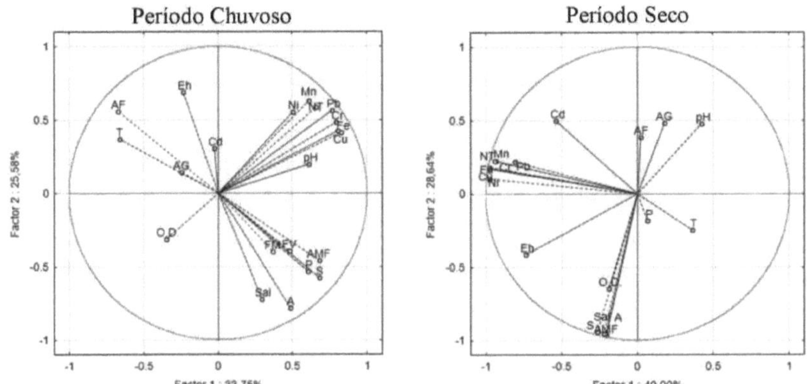

Figure 5.5 - Graphical representation of the Principal Component Analysis of biotic, abiotic, metal and nutrient parameters of the Jequitinhonha River estuarine channel, concerning the campaigns carried out in Nov/2011 and Apr/2012

Note: Metals have been symbolized according to the periodic table, but FV = live foraminifera; FM = dead foraminifera; T = temperature; OD = dissolved oxygen; pH = hydrogen potential; Eh = oxidation-reduction potential; AG = coarse sand; AF = fine sand; AMF = very fine sand; S = silt; A = clay; NT = total nitrogen; MO = organic matter.

5.3.6 Jequitinhonha River estuary mangroves: physico-chemical parameters

With regard to the interstitial waters of the mangrove zones of the Jequitinhonha River, in the first campaign (considered the rainy season), salinity and temperature values oscillated between (0.0ups to 2.0ups and between 25.7°C and 27.0°C) and (5.0ups to 20.0ups and between 26.1°C and 29.3°C, in the second campaign),7°C and 27.0°C) and (5.0ups to 20.0ups and between 26.1°C and 29.3°C, in the second campaign), indicating that the increase in salinity and temperature may be related to the reduction in water volume (AZEVEDO et al., 2011) in the month of sampling (Figure 5.6).

In terms of pH, the mangrove sediment of the Jequitinhonha River was slightly acidic (6.4 to 6.9 in the first campaign) and slightly acidic to neutral (6.1 to 7.4 in the second) (Figure 5.6), which is within the expected range for mangrove areas, since the decomposition of mangrove leaves makes the soil rich in acidic compounds, with pH oscillations between 4.8 and 8.8 (PAULA FILHO et al., 2012).

The Eh values of the sediment were negative at points 1, 4 and 6, characterizing them as a reducing environment. However, the values were positive at points 2, 3 and 5, indicating an oxidizing environment at these sites during the first campaign. However, in the second campaign, the values were negative only at points 2 and 3, giving them a reducing character; and positive values at points 1, 4, 5 and 6, showing a predominantly oxidizing condition during the dry period (Figure 5.6).

5.3.7 Sediment granulometric and nutrient analysis

There was a predominance of silt and very fine sand in both sampling periods, corroborating the low hydrodynamic energy depositional environment characteristic of mangrove areas (Figure 5.6).

In the dry season (sampling in April 2012), there was a reduction in the values of assimilable phosphorus (70.00mg/L to 190.00mg/L in the first campaign and from 60.69mg/L to 120.45mg/L in the second) and total nitrogen (0.093mg/L to 0.279mg/L in the first campaign and from 0.00mg/L to 0.223mg/L in the second) (Figure 5.4). In general, the phosphorus and nitrogen levels were high at the points located close to the downstream, showing contributions from marine waters which tend to dilute their concentrations at the mouth, a situation which becomes more intense during the period, as also observed by Marins et al. (2011) in studies of phosphorus distribution in the sediment of the Jaguaribe river estuary, NE Brazil.

Figure 5.6 - Values of the physicochemical parameters, salinity, temperature, interstitial water and hydrogen potential (pH), oxidation potential (Eh) and granulometric fractions of the sediment of the Jequitinhonha River Mangrove Zone, relative to the Nov/2011 (rainy season) and Apr/2012 (dry season) campaigns.

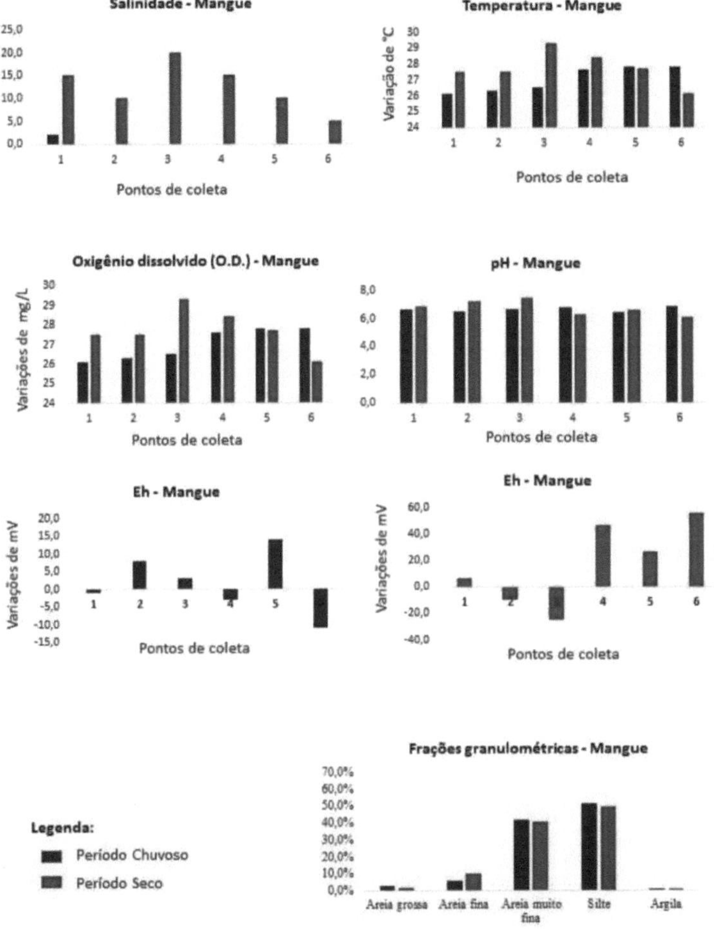

Table 5.4 - Physicochemical, granulometric and nutrient data of the water and sediment of the mangrove zone of the Jequitinhonha River concerning the first (I - Nov/2011) and second (II - Apr/2012) sampling campaigns and limits CONAMA Resolution 357 of March 17, 2005 (Brazil, 2005)

| Samples | Water parameters | | Sediment Analysis | | | | | | | |
	T (°C)	PH	Eh (mV)	GA (%)	PA (%)	MFA (%)	S (%)	A (%)	NT (mg/L	P (mg/L)
MJI1	26,1	6,6	-1	5,3	0,3	25,6	67,4	1,4	0,3	190
MJI2	26,3	6,5	8	8,4	0,1	5,4	55,5	1,3	0,2	137,5
MJI3	26,5	6,6	3	0,6	0,6	51,9	46,2	0,8	0,1	145
MJI4	27,6	6,7	-3	1,6	1,1	49,7	47,1	0,5	0,2	125

	T	pH	Eh	AG	AF	AMF	S	A	P	NT
MJI5	27,8	6,4	14	1,5	0,2	39,2	58,1	1,1	0,1	105
MJI6	27,8	6,9	-11	2,2	2,9	55,5	38,8	0,6	0,1	70
Average	16,21	3,97	1	1,96	0,52	22,73	31,31	0,57	0,1	77,25
MJII1	27,5	6,8	6	3,6	0	19,3	75,6	1,6	0,2	109
MJII2	27,5	7,2	-10	0,7	0,6	49,7	48,9	0,1	0,1	120,5
MJI3	29,3	7,4	-25	0,2	35,4	45,2	19,2	0	0	61,7
MJII4	28,4	6,3	46	1,1	15,6	49,9	33,5	0	0	63,9
MJII5	27,7	6,6	26	1,8	4,5	52,8	40,8	0,1	0,2	60,7
MJII6	26,1	6,1	55	1,4	0	22,8	74,5	1,3	0,2	73
Average	16,65	4,04	9,8	0,88	5,61	23,97	29,25	0,31	0,07	48,88
CONAMA										
SWEET	<40,0	-	-	-	-	-	-	-	-	-
SALINE	-	-	-	-	-	-	-	-	-	-
SALOBRA	-	-	-	-	-	-	-	-	-	-

Legend: MJI1 = Jequitinhonha River mangrove swamp from the first campaign at point 1; T = Temperature; pH = Hydrogen Potential; Eh = Oxirreduction Potential; AG = Coarse Sand; AF = Fine Sand; AMF = Very Fine Sand; S = Silt; A = Clay; P = Phosphorus; NT = Total Nitrogen; CONAMA = National Environment Council.

For comparisons and analysis of the consequences of the high levels of trace metals in the sediments in the study area, we opted to use the limits established by CONAMA Resolution No. 454/2012 (BRASIL, 2012) and the *Canadian Environmental Quality Guidelines* (CCME, 1998), which show low concentrations of Ni, Cr and Zn in the sediment, which means that these metals should not be causing adverse effects on the biota of the Jequitinhonha River channel (Table 5.5). However, these standards do not set limits for Mn and Fe, which makes it impossible to assess the effect of these elements.

While for the Mangrove Zone, comparing the levels of the elements analyzed with the reference limits established by CONAMA Resolution No. 454/2012 (BRAZIL, 2012) and the *Canadian Environmental Quality Guidelines* (CCME, 1998), it was found that only the metal Pb at points 2, 3, 4, 5 and 6 (first campaign) showed values above the limits established by the bodies (Table 5.5). In view of this, it was observed that the high concentrations of lead in the rainy season are possibly related to the drainage of fertilizers and chemical compounds used in agricultural development (ZOURARAH et al., 2009). The high concentrations of the metals Cu (point 2), Pb (point 1) and Cd (point 2) in the dry season are possibly due to the reduction in flow, increasing concentrations in the water and consequently enabling their deposition in the sediments.

5.3.8 Foraminifera fauna

In the first campaign (Nov/2011), 116 individuals were obtained, of which 6 were alive. No fish were found at point 6, despite the high concentrations of trace metals. There were no records of anomalous foreheads at any of the points. In this sampling, 10 species were identified, with *Haplophragmoides wilberti* (37.1%), *Ammonia beccarii* (32.8%), *Quinqueloculina fusca* (12.1%) and *Q. venusta* (5.2%) standing out (Table 5.6). The occurrence of the *A. beccarii* species is due to the fact that it is more tolerant of large changes in salinity and temperature. In addition, it is more tolerant of trace metal contamination, as also shown by Martins et al (2010) in their study on the ecological effects of heavy metals on benthic foraminiferal assemblages in the canals of Aveiro, Portugal. In the second campaign, 143 foraminifera were obtained, of which only 10 were alive, with no anomalous foreheads recorded. Ten species were identified, the main ones being *H. wilberti* (55.9%), *Trochammina inflata* (19.6%) and *Q. fusca* (7.0%) (Table 5.6).

Table 5.5 - Concentrations of metals (mg. Kg -1) in the sediment of the mangrove zone of the Jequitinhonha River, concerning the first (I - Nov/2011) and second (II - Apr/2012) sampling campaigns and reference values from CONAMA Resolution No. 454/2012 (Brazil, 2012) and the *Canadian Environmental Quality Guidelines* (CCME, 1998).

Samples	Ni	Mn	Fe	Cr	Zn	Cu	Pb	Cd
MJI1	8,91	65,58	13769,9	24,99	29,15	11,46	48,16	0,54
MJI2	8,54	83,49	12480,5	20,99	27,52	8,87	**33,39**	0,38
MJI3	8,47	133,34	13524,7	22,02	29,01	8	**36,5**	0,49
MJI4	9,49	73,04	12154,2	22,2	29,42	9,9	**34,77**	0,43
MJI5	6,88	46,73	8312,48	16,22	22,48	7,68	**41,76**	0,27
MJI6	5,96	36,84	6722,17	12,71	18,33	6,18	**33,4**	0,23
Average	8,04	73,17	11160,7	19,855	25,98	8,68	37,99	0,39
MJII1	9,14	67,43	14254,7	29,15	38,88	13,46	**32,75**	0,55
MJII2	11,57	275,57	17617,1	27,52	48,68	**20,44**	48,11	**0,77**
MJII3	6,61	133,92	9398,81	36,78	27,76	7,21	26,12	0,36
MJII4	8,04	51,39	9446,2	18,34	32,95	10,16	54,27	0,29
MJII5	10,17	70,57	12746,8	18,32	38,33	12,28	**38,57**	0,45
MJII6	11,19	54,51	14534,5	31,88	38,46	16,82	**43,53**	0,59
Average	9,45	108,89	12999,7	26,99	37,51	13,39	40,55	0,5
LD	0,00415	0,00111	0,02586	0,01632	0,00651	0,00336	1,49	0,00095
CONAMA N1	20,9	n.a.	n.a.	81	150	34	46,7	1,2
CONAMA N2	51,6	n.a.	n.a.	370	410	270	218	7,2
CEQG ISQG	n.a.	n.a.	n.a.	52,3	124	18,7	30,2	0,7
CEQG PEL	n.a.	n.a.	n.a.	160	271	108	112	4,2

Legend: LOD = Limit of Detection; N1 and ISQG= threshold below which there is a lower probability of adverse effects on biota; N2 and PEL= threshold above which there is a higher probability of adverse effects on biota; n.a. = not determined.

Thus, the increase in foraminifera is due to the significant increase in salinity, and the same was found by Martins et al (2010). In addition, the predominance of the species *H. wilberti*, in both sampling periods, suggests the presence of environments composed of fine sedimentation and rich in nutrients, typical of mangrove regions (GÓMES; BERNAL, 2011; MACHADO et al., 2012). The genus *Quinqueloculina* is typical of marine environments, showing a high number of specimens during the first campaign, which had the highest salinities. The low number of specimens recorded during the second campaign indicates a mixohaline environment, as was also observed by Rodrigues et al (2003) in studies monitoring foraminifera in the Bertioga channel - SP, Brazil.

The richness values were higher in the first sampling period (0.4 to 2.0 in the rainy season and 0.5 to 1.7 in the dry season - Table 5.6), but in both campaigns, the evenness indices were above 0.5, indicating a lack of dominance, corroborating the statistical studies on foraminifer assemblages carried out by Clarke and Warwick (2001). Diversity was lower in the dry season (0.3 to 1.4 in the first campaign and 0.3 to 1.3 in the second) (Table 5.6). However, the diversity values were considered high when compared to the data obtained by Semensatto-Jr (et al., 2009) in a study carried out in a mangrove environment located to the north of Ilha do Cardoso, Cananéia-Iguape Bay, SP (0.2 to 0.6).

Although there was no dominance, the higher richness indices in the dry season are due to the tolerance of some species to contamination by trace metals (GÓMES; BERNAL, 2011).

5.3.9 Taphonomy of the foreheads

During the rainy sampling period, the majority of the foreheads were white or colorless, 72.27%, and during the dry period, the majority were brown, 60.00%. As for wear, 51.26% and 40.60% of the foreheads were abraded and normal (Figure 5.7).

Thus, the dominance of white tests results from the rapid addition of new tests to the sediment, as verified by Machado et al (2012) in taphonomic studies with foraminifer assemblages in the municipality of Conde, Bahia (Figure 5.7). However, there was also a high percentage of abraded foreheads, which indicate high-energy conditions. Considering that most of the abrased specimens were of the *A. beccarii* species and that this species can also be found in the estuarine channel of the Jequitinhonha River (Figure 5.4 and Tables 5.3), it is possible to infer that some of the foreheads were transported into the mangrove due to the speed of the currents in the region, as observed by Rodrigues et al (2003) in similar studies carried out in the Bertioga - SP channel. In the second campaign, there was a predominance of brown shells, suggesting a slow rate of sedimentation, caused by intense but not rapid erosion or bioturbation (MACHADO et al., 2012), which is confirmed by the predominance of normal shells.

Table 5.6 - Absolute (N) and relative (AR) abundances and ecological indices of foraminifer species recorded in the mangrove zone of the Jequitinhonha River estuary during the first (I - Nov/2011) and second (II - Apr/2012) sampling campaigns

Jequitinhonha River mangrove swamp	JI1	JI2	JI3	JI4	JI5	JI6	N	AR	JII1	JII2	JII3	JII4	JII5	JII6	N	AR
Ammonia beccarii	0	30	7	1	0	0	38	**33**	2	1	2	0	0	0	5	3,5
Ammonia parkinsoniana	0	1	0	0	0	0	1	0,9	0	0	0	0	0	0	0	0
Bolivina laevigata	0	0	0	0	0	0	0	0	3	0	0	0	0	0	3	2,1
Brizalina alata	0	0	0	0	0	0	0	0	0	3	2	0	0	0	5	3,5
Elphidium excavatum	0	0	0	0	0	0	0	0	0	0	1	0	0	0	1	0,7
Elphidium sagrum	0	3	0	0	0	0	3	2,6	0	1	0	0	0	0	1	0,7
Haplophragmoides wilberti	2	37	2	2	0	0	43	**37**	10	6	15	44	4	1	80	**55,9**

Nonion grateloupi	0	0	2	0	0	0	2	1,7	0	0	0	0	0	0	0	0
Quinqueloculina fusca	14	0	0	0	0	0	14	**12**	5	0	3	2	0	0	10	7
Quinqueloculina seminula	0	0	4	0	0	0	4	3,5	0	5	0	0	0	0	5	3,5
Quinqueloculina venusta	0	3	3	0	0	0	6	**5,2**	0	0	1	0	0	0	1	0,7
Textularia aglutinans	0	0	2	0	0	0	2	1,7	4	0	0	0	0	0	4	2,8
Trochammina inflata	0	0	1	0	2	0	3	2,6	15	3	7	3	0	0	28	**19,6**
Total	16	74	21	3	2	0	**116**	100	39	19	31	49	4	1	**143**	100
No. of species	2	5	7	2	1	0	-	-	6	6	7	3	1	1	-	-
Wealth (Margalef)	0,4	0,9	2	0,9	0	****	-	-	1,4	1,7	1,7	0,5	0	****	-	-
Evenness (Pielou)	0,5	0,6	0,9	0,9	****	****	-	-	0,9	0,9	0,8	0,4	****	****	-	-
Diversity (S-Wiener)	0,3	0,9	1,4	0,4	0	****	-	-	1,3	1,3	1,2	0,3	0	0	-	-

Legend: MPI1 = Jequitinhonha River from the first campaign at point 1; N = Total; RA = Relative Abundance; Diversity (S - Wiener) = Diversity (Shannon-Wiener).

Figure 5.7 - Percentages of the types of coloration and wear of foraminifer species in the mangrove zone of the Jequitinhonha River, referring to the Nov/2011 and Apr/2012 campaigns.

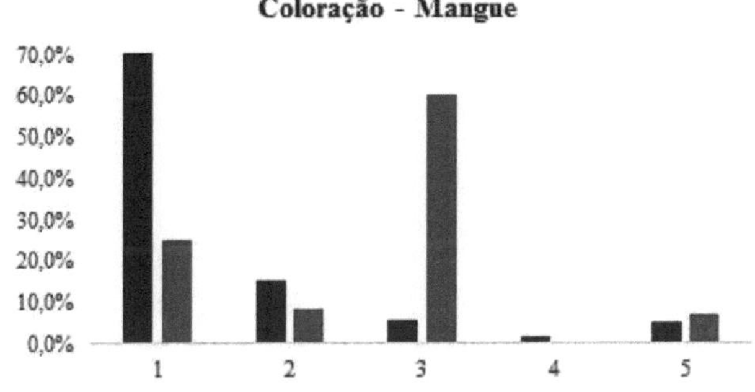

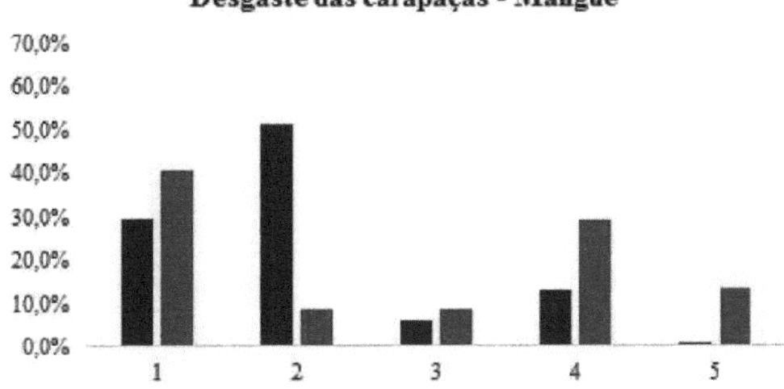

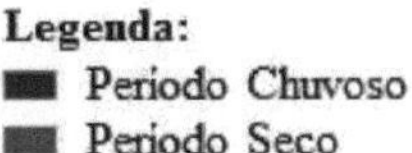

5.3. 10Distribution of foraminiferal heads

The results of the correlation analysis for both campaigns indicate that the distribution of living individuals is positively correlated with the parameters temperature, pH and the fractions fine sand, very fine sand and silt (Figure 5.8), suggesting that these organisms prefer fine-grained environments, which favor the supply of nutrients. The foreheads of the dead individuals also showed a positive correlation for both campaigns with the silt and clay fractions and the trace metal Pb (Figure 5.8). This is because, after the death of these organisms, the foreheads are deposited in environments with lower hydrodynamic energy, where fine-grained sediments predominate, which also favors greater adsorption of trace metals.

Figure 5.8 - Graphical representation of the Principal Component Analysis, based on biotic, abiotic, metal and nutrient parameters in the sediment of the mangrove zone of the Jequitinhonha River, concerning the campaigns carried out in Nov/2011 and Apr/2012

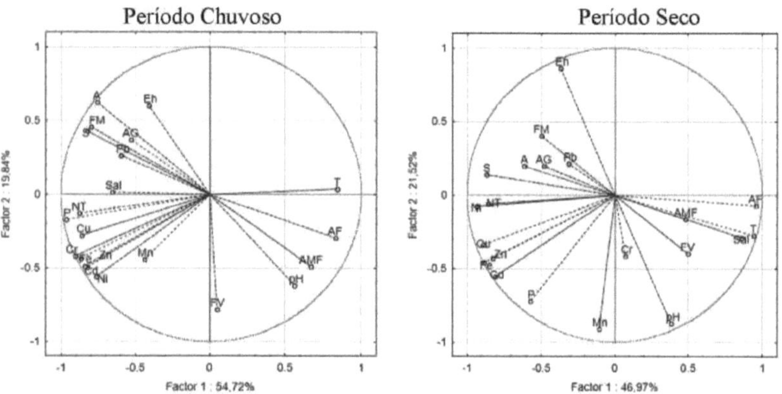

Note: Metals have been symbolized according to the periodic table, but FV = live foraminifera; FM = dead foraminifera; T = temperature; OD = dissolved oxygen; pH = hydrogen potential; Eh = oxidation-reduction potential; AG = coarse sand; AF = fine sand; AMF = very fine sand; S = silt; A = clay; NT = total nitrogen; MO = organic matter.

5.4 CONCLUSIONS

The estuarine channel of the Jequitinhonha River is made up of fresh water, at points 1 to 4 (first campaign) and points 1 and 3 to 6 (second campaign); brackish water, only at point 2 (second campaign), and slightly oxygenated. In addition, salinity and temperature values were lower in November 2011 than in April 2012 due to an anomaly in the region's rainfall regime, so that, contrary to expectations, the first campaign was considered to have been carried out in the "rainy season" and the second in the "dry season".

The sediment is slightly acidic to neutral and markedly reducing during the rainy season. Fine sand was the predominant granulometric fraction, confirming the expectation of a condition of lower hydrodynamic energy in the channel.

The concentrations of assimilable phosphorus, organic matter, organic carbon, nickel, iron and chromium in the canal probably originate from the drainage of agricultural areas upstream, mainly during the rainy season, while nitrogen probably results from urban and industrial effluents which, like cadmium concentrations, are more concentrated during the dry season.

In the Jeqtuitinhonha river channel, 281 individuals were obtained during the rainy season (Nov/2011), with *Trochammina inflata* (37.0%), *Haplophragmoides wilberti* (22.8%) *and Ammonia beccarii* (18.5%) standing out as the main species. The *A. beccarii* species is the most tolerant of major changes in temperature and salinity and also has a preference for fine sedimentation with nutrient inputs. In the dry season, the lack of foraminifera was due to the low salinity values, which made it impossible for even the most resistant species to survive. Despite this, there was no dominance of species, although the richness and diversity values were low compared to other estuaries.

During the rainy season, most of the foreheads were white or colorless, 63.70%, or yellow, 25.62%. As for wear, 44.48% of the foreheads were normal, but 29.89% showed signs of abrasion.

The predominance of white foreheads is indicative of a very rapid deposition rate, with a lot of new material being added to the sediment, especially during the dry season, which is corroborated by the dominance of normal foreheads.

In the first campaign, the distribution of living and dead individuals was positively correlated with the very fine sand, silt, clay and assimilable phosphorus fractions, indicating that fine sediment has a higher organic content, showing a direct association between sediment texture and benthic foraminifera. On the other hand, the foreheads of dead individuals show a positive correlation with the clay, silt and pH fractions, given that, after the death of these organisms, the foreheads are usually deposited in environments with lower hydrodynamic energy, where, consequently, fine-grained sediments predominate.

For the second campaign, the absence of foraminifera is possibly due to the low salinity values and contamination by trace metals, compromising the survival of the species and making it impossible to carry out an analysis of the correlation between the organisms and the abiotic parameters.

For the mangrove zone of the Jequitinhonha River estuarine channel, the increase in salinity and temperature values during the second campaign may be related to the reduction in water volume.

Its sediment is slightly acidic to neutral and predominantly oxidizing in the dry season. The very fine sand and silt fractions were the predominant granulometric compositions, confirming a depositional environment with lower hydrodynamic energy.

The concentrations of assimilable phosphorus and nitrogen come from marine waters which tend to dilute their concentrations at the mouth, a situation which becomes more intense during the rainy season. The levels of cadmium, chromium, lead, iron and zinc are probably the result of anthropogenic sources, such as domestic effluents and agricultural drainage from shrimp farming activities, while nickel is probably the result of mining activities and industrial waste in the region.

In the mangrove zone, 116 individuals were obtained. No testes were found at point 6 and despite the high concentrations of trace metals, no anomalous testes were recorded at any of the points. In this sampling, 10 species were identified, with *Haplophragmoides wilberti* (37.1%), *Ammonia beccarii* (32.8%), *Quinqueloculina fusca* (12.1%) and *Q. venusta* (5.2%) standing out. The occurrence of these species is due to the fact that they are more tolerant of large changes in salinity and temperature. In addition, the species *H. wilberti* and *A. beccarii* are the most tolerant to contamination by trace metals. In the second campaign, 143 foraminifera were obtained, with no anomalous foreheads recorded. Ten species were identified, the predominant ones being: *H. wilberti* (55.9%), *Trochammina inflata* (19.6%) and *Q. fusca* (7.0%). The increase in the number of foreheads is due to the significant increase in salinity during the second campaign.

During the rainy sampling period, the majority of the testes were white or colorless, 72.27%, and during the dry period, the majority were brown, 60.00%. As for wear and tear, the percentage of abraded testes was 51.26%, while normal testes accounted for 40.60%.

Thus, the dominance of white cores results from the rapid addition of new cores to the sediment. However, there was also a high percentage of abraded foreheads, which indicate high-energy conditions. Considering that most of the scorched specimens were of the *A. beccarii* species and that it also lives in the estuarine channel of the Jequitinhonha River, it is possible that some of the foreheads were transported into the mangrove. In the second campaign, there was a predominance of brown shells, suggesting a slow rate of sedimentation, caused by intense but not rapid erosion or bioturbation, which is confirmed by the predominance of normal shells.

The correlation analyses indicate that the distribution of live individuals for both campaigns was positively correlated with the parameters temperature, pH and the fractions fine sand, very fine sand and silt, suggesting that these organisms prefer fine-grained environments, which favor the supply of nutrients. The foreheads of the dead individuals also showed a positive correlation with the silt and clay fractions and the trace metal Pb for both campaigns. This is because, after the death of these organisms, the foreheads are deposited in environments with lower hydrodynamic energy, where fine-grained sediments predominate, which also favors greater adsorption of trace metals.

5.5 REFERENCES

AZEVEDO, J.S.; BRAGA, E.S.; FAVARO, D.T.; PERRETTI, A.R.; REZENDE, C.E.; SOUZA, C.M.M. Total mercury in sediments and in Brazilian Ariidae catfish from two estuaries under different anthropogenic influence. **Marine Pollution Bulletin**, v.62, n.12, p.2724-2731, dec 2011.

BOUILLON, A.V.; BORGES, E.; CATANEDA-MOYA, K.; DIELE, T.; DITTMAR, N.C.; DUKE, E.; KRISTENSEN, S.Y.; LEE, C.; MARCHAND, J.J.; MIDDELBURG, V.H.; RIVERA-MONROY, T.J.; SMITH, R.R.T. Mangrove production and carbon sinks: a version of global budget estimate Global Biogeochem. **Global Biogeochemical Cycles,** v.22, n.2, p.112, may. 2008.

BRAZIL. Ministry of the Environment. National Environment Council, CONAMA. **CONAMA Resolution No. 357,** March 17, 2005. Available at: <http://www.mma.gov.br/port/conama/legiabre.cfm?codlegi=459>. Accessed on: March 13, 2014.

CELINO, J.J.; ESCOBAR, N.F.C.; HADLICH, G.M.; NASCIMENTO, R.A.; QUEIROZ, A.F.S. Geochemistry of surface water in the lower reaches of the Una, Pardo and Jequitinhonha rivers, southern Bahia. In: CELINO, J.J.; HADLICH, G.M.; QUEIROZ, A.F.S.; OLIVEIRA, O.M.C. (Orgs.). **Evaluation of coastal environments in the southern region of Bahia:** geochemistry, oil and society. Salvador: EDUFBA, 2014. p.63-76.

COSENTINO, D.; BERTINI, A.; CIPOLLARI, P.; FLORINDO, F.; GLIOZZI, E.; GROSSI, F.; LO MASTRO, S.; SPROVIERI, M. Orbitally forced paleoenvironmental and paleoclimate changes in the late postevaporitic Messinian of the central Mediterranean Basin. **The Geological Society of America Bulletin,** v.124, n.3-4, p.499-516, mar. 2012.

CRUZ, F.C. **Trace elements in mangrove substrate from the municipalities of Una, Canavieiras and Belmonte, Bahia.** 2012. 110f. Dissertation (Master's Degree in Geochemistry) - Institute of Geosciences, Federal University of Bahia, Salvador, 2012.

CPRM - Mineral Resources Research Company. Project to Register Groundwater Supply Sources - Jequitinhonha Valley. **Diagnosis of the Municipality of Belmonte - BA.** Minas Gerais: p.157, 2005.

CARVALHO, P.V.; SANTOS, P.J.; BOTTER-CARVALHO, M.L. Assessing the severity of disturbance for intertidal and subtidal macrobenthos: the phylum-level meta-analysis approach in tropical estuarine sites of northeastern Brazil. **Marine Pollution Bulletin,** v.60, n.6, p.87387, jun. 2010.

CLARKE, K.R.; WARWICK, R.M. **Change in marine communities:** an Approach to Statistical Analysis and Interpretation. 2nded. [s.l.]: Plymouth Marine Laboratory, 2001. 172p.

DONNICI, S.; SERANDREI-BARBERO, R.; BONARDI, M.; SPERLE, M. Benthic foraminifera as proxies of pollution: The case of Guanabara Bay (Brazil). **Marine Pollution Bulletin**, v.64, n.10, p.2015-2028, oct. 2012.

DUQUET, M. **Ciências da vida:** glossàrio de ecologia fundamental. Porto: Porto Editora, 2007. 128p.

DAJOZ, R. **General ecology**. 2.ed. Petrópolis: Vozes, 1983. 472p.

ESCOBAR, N.F.C.; CELINO, J.J.; NASCIMENTO, R.A. Metals in surface water, suspended particulate matter and bottom sediment in the lower reaches of the Una, Pardo and Jequitinhonha rivers. In: CELINO, J.J.; HADLICH, G.M.; QUEIROZ, A.F.S.; OLIVEIRA, O.M.C. (Orgs.). **Evaluation of coastal environments in the southern region of Bahia:** geochemistry, oil and society. Salvador: EDUFBA, 2014. 77-98p.

ESCOBAR, N.F.C. **Geochemistry of surface water and bottom sediment in the lower reaches of the Una, Pardo and Jequitinhonha rivers, Southern Bahia, Brazil.** 2013. 107f. Dissertation (Master's Degree in Geochemistry) - Institute of Geosciences, Federal University of Bahia, Salvador, 2013.

FRONTALINI, F.; COCCIONI, R. Benthic foraminiferal for heavy metal pollution monitoring: A case study from the central Adriatic Sea coast of Italy. **Estuarine Coastal and Shelf Science**, v.76, n.2, p.404-417, jan. 2008.

FERREIRA, V.O. **Paisagem, Recursos Hidricos e Desenvolvimento Econòmico na Bacia do Rio Jequitinhonha, MG.** 2007. 291f Thesis (Doctorate in Geography) - Federal University of Minas Gerais, Belo Horizonte, 2007.

GÓMES, E.; BERNAL, G. Influence of the environmental characteristics of mangrove forests on recent benthic foraminifera in the Gulf of Urabà, Colombian Caribbean. **Marine Sciences**, v.39, n.1, p. 69-82, 2013.

GUERRERO, P. Vale do Jequitinhonha: The region and its contrasts. **Revista Discente Expressoes Geogràficas**, v.5, n.5, p.81-100, May 2009.

IBGE - Brazilian Institute of Geography and Statistics. **Vegetation maps of Brazil.** Rio de Janeiro: IBGE, 2004. 1 map, color. Scale 1:5,000,000. Available at <ftp://ftp.ibge.gov.br/Cartas_e_Mapas/Mapas_Murais/>. Accessed on: April 19, 2015.

IBGE - Brazilian Institute of Geography and Statistics. **Environmental diagnosis of the Jequitinhonha river basin** - general guidelines for land use planning. Fundaçao Instituto Brasileiro de Geografia e Estatistica IBGE, Diretoria de Geociências. Salvador, 1997. Available at:

<http://biblioteca.ibge.gov.br>. Accessed on: April 19, 2014.

LAMPARELLI, C.C. Commissioning and monitoring challenges regarding ocean outfalls: Sao Paulo State experience. In: Lamparelli, C.C.; Ortiz, J.P. (Eds.). **Submarine outfalls:** design, compliance and environmental monitoring. Sao Paulo: Secretaria do Meio Ambiente, 2007. p.11-23.

MACHADO, A.J.; ARAÙJO, T.M.F.; ARAÙJO, H.A.B.; FIGUEIREDO, S.M.C. Bathymetric and taphonomic analysis of the foraminiferal microfauna of the continental shelf and slope of the Municipality of Conde, Bahia. **Cadernos de Geociências,** v.9, n.2, p. 157-172, nov. 2012.

MARTINS, R.V.; FILHO, P.F.J.; ESCHRIQUE, S.A.; LACERDA, L.D. Anthropogenic sources and distribution of phosphorus in sediments from the Jaguaribe River estuary, NE, Brazil. **Brazilian Journal of Biology,** v.71, n.3, p. 673-678, aug. 2011.

MARTINS, M.V.A.; DA SILVA, E.F.; SEQUEIRA, C.; ROCHA, F.; DUARTE, A.C. Evaluation of the ecological effects of heavy metals on the assemblages of benthic foraminifera of the canals of Aveiro (Portugal). **Estuarine, Coastal and Shelf Science (Print),** v.87, n.2, p.293-304, apr. 2010

MURRAY J. **Ecology and applications of benthic foraminifera.** Englan: Cambridge University Press, 2006. 426p.

ONOFRE, C.R.E.; CELINO, J.J.; NANO, R.M.W.; QUEIROZ, A.F.S. Bioavailability of trace metals in mangrove sediments from the northern portion of Todos os Santos Bay, Bahia, Brazil. **Revista de Biologia e Ciências da Terra,** v.7, n.2, 2.Sem. 2007.

PAULA FILHO, P.F.J.; DE MOURA, M.C.S.; MARINS, R.V. Phosphorus Geochemical Fractioning in Water and Sediment from Corrente River, Catchment, ParnaibaZPI. **Revista Virtual de Quimica,** v.4, n.6, p.623-640, nov. 2012.

PERH-BA. Bahia State Water Resources Plan. **Diagnosis and Regionalization.** Salvador, 2003 (Final report for stage I). Available at: <http://biblioteca.inga.ba.gov.br>. Accessed: Aug. 18, 2012.

QUEIROZ, A.F.S.; OLIVEIRA, O.M.C. **Geoenvironmental diagnosis of mangrove areas and development of technological processes applicable to the remediation of these areas:** subsidies for an impact prevention program in areas with potential for oil activities in the southern coastal region of the State of Bahia (PETROTECMANGUE-BASUL). Salvador: EDUFBA, 2013. 148p. (Technical report).

RODRIGUES, A.R.; EICHLER, P.P.B.; EICHLER, B.B. Utilization of foraminifera in the monitoring of the Bertioga channel (SP, Brazil). **Atlântica, Rio Grande,** v.25, n.1, p.35-51, 2003.

SAMPAIO, N. E VARGAS, M.A.M. The landscape of the Pardo river unveiled by riverine community in southwestern Bahia: Talks between the perceived and lived. **Revista Eletrônica Ateliê Geogràfico,** v.4, n.4, p.147-177, dec. 2010.

SIQUEIRA, G.W.; LIMA, W.N.; MENDES, A.S.; APRIL, F.M.; BRAGA, E.S.; MAHIQUES, M.M. Evolution of the environmental impact caused by organic matter, mercury and arsenic in the bottom sediments of the Santos estuarine system. **Geochimica Brasil,** v.18, n.1, p.054-063, 2004.

TEODORO, A.C.; DULEBA, W.; GUBITOSO, S.; PRADA, S.M.; LAMPARELLI, C.C.; BEVILACQUA, J.E. Analysis of foraminifera assemblages and sediment geochemical properties to characterize the environment near Araçâ and Saco da Capela domestic sewage submarine outfalls of Sao Sebastiao Channel, Sao Paulo State, Brazil. **Marine Pollution Bulletin,** v.60, p.536-553, 2010.

TEODORO, A.C.; DULEBA, W.; LAMPARELLI, C.C. Foraminiferal associations and textural composition of the region near the Cigarras domestic sewage submarine outfall, Sao Sebastiao Channel, SP, Brazil. **Pesquisas em Geociências,** v.36, n.1, p.467-475, jan./abr. 2009.

VEIGA, J.E. Indicadores de sustentabilidade. **Estudos Avançados,** v.24, n.68, p.39-52, 2010.

ZOURARAH, B.; MAANAM, M.; ROBIN, M. Sedimentary records of anthropogenic contribution to heavy metal content in Oum Er Bia estuary (Morocco). **Environ Chem Lett.,** v.7, p.67-78, 2009.

6 GENERAL CONCLUSION

Among the sediment geochemical and biotic data from this research, salinity and temperature values were lower in November 2011 than in April 2012 due to an anomaly in the rainfall regime of the regions under study, also showing a predominantly reducing environment in the three regions during the rainy season.

The concentrations of assimilable phosphorus, organic matter, organic carbon and the metals nickel, chromium and iron were similar in the three estuaries in terms of their origin, perhaps coming from the drainage of agricultural areas upstream, especially during the rainy season, while nitrogen probably comes from urban and industrial effluents. The absence and significant reduction of foraminifera is due to the reduction in salinity and also high concentrations of trace metals, with the species *Elphidium excavatum, Ammonia tepida* and *A. beccarii standing* out, in that order, as the most tolerant to contamination by non-essential metals in the three study areas. While the occurrences of the species *H. wilberti, T. squamata* and *T. inflata* indicate estuarine environments with low hydrodynamic energy, as observed in the estuaries of the Pardo and Jequitinhonha rivers, and a tendency towards hyposalinity, a peculiarity presented in the three estuarine channels, given the predominance of brackish waters.

The sedimentary characteristics of the regions studied were similar in terms of composition, with slightly acidic to neutral sediments and predominantly oxidizing sediments during the rainy season. In granulometric terms, the very fine sand and silt fractions were the predominant compositions, confirming a depositional environment of lower hydrodynamic energy for the three mangrove zones of the estuarine channels studied.

The concentrations of assimilable phosphorus in the mangrove zones of the Una and Pardo rivers are possibly due to anthropogenic sources, i.e. domestic effluents or carciniculture close upstream. Nitrogen, on the other hand, probably comes from natural sources, such as assimilation and consumption by phytoplankton. Similarly, the concentrations of assimilable phosphorus and nitrogen in the mangrove zone of the Jequitinhonha River also appear to come from natural sources, such as marine waters that have tended to dilute their concentrations at the mouth, a situation that becomes more intense during the rainy season.

The trace metals studied in the three regions were cadmium, chromium, lead and zinc, which seem to have originated from agricultural drainage, especially during the rainy season, and nickel, which is the result of mining activities and industrial waste.

In all three areas, significant reductions in foraminifera were recorded during the dry season, due to the high levels of trace metals, which enabled the survival of the more tolerant species, such as *A.*

tepida and *E. excavatum,* which are dominant in the three mangrove zones. In addition, the presence of the species *Haplophragmoides wilberti, Trochammina inflata* and *T. squamata* was recorded in the regions studied, as these species have a preference for fine sedimentation environments, as they have a higher organic content in their composition.

The taphonomic results revealed similarities and peculiarities in the three channels and mangrove zones of the estuaries studied. A predominance of white tests was observed, indicating a very rapid rate of deposition, with a lot of new material being added to the sediment, especially during the dry season, which is corroborated by the dominance of normal tests. In addition, there was no species dominance in any of the environments studied, although the richness and diversity values were low when compared to other studies.

REFERENCES

AGUIAR, P.C.B.; BRUNO, N.L.; MOREAU, A.M.S.S.; FONTES, E.O. Geographical occupation and evolution of the configuration of the territory of the municipality of Canavieiras, Bahia. **Revista de Geografia, Meio Ambiente e Ensino,** v.3, n.2, p.49-67, 2°Sem. 2012.

AGUIAR, P.C.B.; MOREAU, A.S.S.; FONTES, E.O. Impacts on the environmental dynamics of the municipality of Canavieiras (BA) with RESEX as a factor of influence. **Revista de Geografia, Meio Ambiente e Ensino,** v.2, n.1, p. 61-78, 1°Sem. 2011.

ARMSTRONG. H.; BRASIER, M. **Microfossils.** 2nd ed. Oxford: Blackwell Publishing, 2005. 296p.

ARAÙJO, T.M.F. **Estudo da microfauna de foraminiferos do sedimento da superfície a da subsuperfície da plataforma e do talude continental da regiào norte do estado da Bahia (Salvador à Barra do Itariri).** 2004. 527f. Thesis (Doctorate in Geology) - Institute of Geosciences, Federal University of Bahia, Salvador, 2004.

ANDRADE, E.J. **Distribution of recent foraminifera in the carbonate/siliclastic transition in the Praia do Forte region, North Coast of the State of Bahia. Salvador.** 1997.   111f. Dissertation (Master's Degree in Geology) - Institute of Geosciences,

Federal University of Bahia, Salvador, 1997.

ALVE, E. Benthic foraminiferal responses to estuarine pollution: a review. **The Journal of Foraminiferal Research,** v.25, p.190-203, 1995.

ALVE, E. Benthic foraminifera in sediment cores reflecting heavy metal pollution in Sorfjord, Western Norway. **The Journal of Foraminiferal Research,** v.25, n.3, p.190-203, jul. 1995.

ALMASI, M.N. **Ecology and color variation of benthic foraminifera in Barnes Sound,**

Northeast Florida Bay. 1978, 144f. (ecology in Masters) - University of Miami, Miami, 1978.

BRITO, E.M.; DURAN, R.; GUYONEAUD, R.; GONI-URRIZA, M.; GARCIA, O.T.; CRAPEZ, M.A. A case study of in situ oil contamination in a mangrove swamp (Rio de Janeiro, Brazil). **Marine Pollution Bulletin,** v.58, p.418-423, 2009.

BRAZIL. Ministry of the Environment. National Environment Council, CONAMA. **CONAMA Resolutions No. 401**, November 4, 2008. Available at:

< http://www.mma.gov.br/port/conama/legislacao> Accessed on: 19 Jan. 2014.

BOEGER, M.R.T.; MUSCHNER, V.C.; PIE, M.R.; OSTRENSKY, A.; BOEGER, W.A. Postglacial North-South expansion of population of Rhicophora *mangle* (Rhizophoraceae) along the Brazilian coast revealed by microsatellite analysis. **American Journal of Botany,** v.98, n.6, p.1031-1039, 2011.

BARBOSA, C.F.; OLIVEIRA-SILVA, P.; FERREIRA, B.P.; SEOANE, J.C.S.; ALMEIDA, C.M.; MARINHO, L.S.V. Application of the FORAM index to assess the health of coral reefs in the parrachos de maracajaù (RN). In: II Congresso Brasileiro de Oceanografia, 2005, Vitória. **2 Congresso Brasileiro de Oceanografia.** p.1-4, 2005.

BARRETO, A.C. **Qualidade do solo de uma microbacia do rio Una (Aliança) - BA sob diferentes usos da Terra.** 2005. 76f. Dissertation (Master's Degree in Soil Sciences) - Federal Rural University of Pernambuco, Recife, 2005.

BITTENCOURT, A.C.S.P.; DOMINGUEZ, J.M.L.; MARTIN, L.; SILVA, I.R. Patterns of Sediment Dispersion Coastwise the State of Bahia - Brazil, 72., 2000, Bahia: **Anais da Academia Brasileira de Ciências,** . v.72, n.2, p.271-287, 2000.

BRONNIMANN, P.; DIAS-BRITO, D. New Iituolacea (Protista, Foraminifera) from shallow waters of the Brazilian Shelf. **The Journal of Foraminiferal Reasearch,** v.12, n.1, p.13-23, 1982.

BOLTOVSKOY, E.; GIUSSANI, G.; WATANABE, S.; WRIGHT, R.C. **Atlas of benthic shelf foraminifera of the southwest atlantic.** Netherlands: The Hague, 1980, 133Pp.

BRONNIMANN, P.; WHITTAKER, J.E.; ZANINETTI, L. Brackish water foraminifera from mangrove sediments of southwestern Viti Levu, Fiji Islands, Southwest Pacific. **Revué de Paléobiologie,** v.11, n.1, p.13-65, 1992.

BOLTOVSKOY, E.; VIDARTE, M. Foraminifera of the mangrove zone of Guayaquil (Ecuador). **Revista del Museo Argentino de Ciencias Naturales,** v.5, n.3, p.31-49, 1977.

BOLTOVSKOY, E.; WRIGHT, R. **Recent foraminifera.** [s.l.]: Publishers The Hague, 1976. 515p.

BOLTOVSKOY, E. **Los foraminiferos recientes.** Argentina: EUDEBA, University of Buenos Aires, 1965. 510p.

CHÂTELET, E.A.; BOUT-ROUMAZEILLES, V.; RIBOULLEAU, A.; TRENTESAUX, A. Sediment (grain size and clay mineralogy) and organic matter quality control on living foraminifera. **Revue de Micropaléontologie**, v.52, n.1, p.75-84, jan./mar. 2009.

CRUZ, F.C. **Background levels of trace elements in mangrove substrate from the municipalities of Una, Canavieiras and Belmonte, Bahia.** 2013. 93f Dissertation (Master's Degree in Geochemistry) - Institute of Geosciences, Federal University of Bahia, Salvador, 2013.

COSTA, A.B.; NOVOTNY, E.H.; BLOISE. A.C.; DE AZEVEDO, E.R.; BONAGAMBA, T. J.; ZUCCHI. M.R; SANTOS. V.L.C.S.; AZEVEDO, A.E.G. Characterization of organic matter in sediment cores oh the Todos os Santos Bay, Bahia, Brazil, by Elemental analysis and[13] C NMR. **Marine Pollution Bulletin,** v.62, n.8, p.1883-1890, 2011.

CRONA, B.I.; RONNBACK, P.; JIDDAWI, N.; OCHIEWO, J.; MAGHIMBI, S.; BANDEIRA, S. Murky water: Analyzing risk perception and stakeholder vulnerability related to sewage impacts in mangroves of East Africa. **Global Environmental Change,** v.19, p.227239, 2009.

CPRM - Mineral Resources Research Company. Project to register underground water supply sources in the Jequitinhonha Valley. **Diagnosis of the municipality of Belmonte - BA.** Minas Gerais, 2005.

CULVER, S.J. Foraminifera. In: LIPPS, H. J (Orgs.). **Fossil prokaryotes and protests.** Oxford: Blackwell Scientific Publications, 1993. 203-246p.

COTTEY T.L.; HALLOCK, P. Test surface degradation in Archaias angulatus. **The Journal of Foraminiferal Research,** v.18, n.3, p.187-202, 1988.

CLOSS, D. Foraminifera and tecamebae from Lagoa dos Patos (RS). **Boletim da Escola de Geologia de Porto Alegre,** v.11, p.1-130, 1962.

CLOOS, D.; BARBERENA, M.C. Recent foraminifera from Praia da Barra (Salvador, Bahia). **Geological School of Porto Alegre,** v.6, p.1-50, 1960.

DELGARD, M.; DEFLANDRE, B.; METZGER, E.; NUZZIO, D.; CAPO, S.; MOURET, A.; ANSCHUTZ, P. In situ study of short-term variations of redox species chemistry in intertidal permeable sediments of the Arcachon lagoon. **Hydrobiologia,** v.699, n.1, p.69-84, dec. 2012

DE PAULA, F.C.F.; SILVA, D.M.L.; SOUZA, C.M. Typologias hidroquimicas das bacias hidrogràficas do leste da Bahia, **Revista Virtual Quimica.,** v.4, n.4, p.365-373, abr. 2012.

DONNICI, S.; SERANDREI-BARBERO, R.; BONARDI, M.; SPERLE, M. Benthic foraminifera as proxies of pollution: The case of Guanabara Bay (Brazil). **Marine Pollution Bulletin,** v.64, n.10, p.2015-2028, oct. 2012.

DIAS, H.M.; SOARES, M.G.; NEFFA, E. Socio-environmental conflicts: the case of shrimp farming in the Caravelas - Nova Viçosa/Bahia-Brazil estuarine complex. **Revista Ambiente e Sociedade,** v.15, n.1, p. 111-130, 2012.

D'AQUINO, C.A.; NETO, J.S.A.; BARRETO, G.A.M.; SCHETTINI, C.A.F. Oceanographic characterization and suspended sediment transport in the Mampituba River estuary, SC. **Revista Brasileira de Geofisica,** v.29, n.2, p.217-230, 2011.

DIEZ, M.; SIMON, M.; MARTIN, F.; DORRONSORO, C.; GARCIA, I.; VAN GESTEL, C. A.M. Ambient trace element background concentrations in soils and their use in risk assessment. **Science of the Total Environment,** v.407, p.4622-4632, 2009.

DIZ, P.; FRANCÉS, G.; ROSÓN, G. Effects of Contrasting Upwelling-Downelling on Benthic Foraminiferal Distribution in the Ria de Vigo (NW Spain). **Journal of Marine Systems,** v.60, p.1-18, 2006.

DEBENAY, J. P., GUIRAL, D.; PARRA, M. The Case of Foraminiferal Assemblages in French Guiana. **Estuarine, Coastal and Shelf Science,** v.55, p.509-533, 2002.

DEBENAY, J.P.; DULEBA, W.; BONETTI, C.; MELO-E-SOUZA, S.H.; EICHLER, B.B. Pararotalia cananeiaensis n. sp., indicator of marine influence and water circulation in Brazilian coastal and paralic environments. **The Journal of Foraminiferal Research,** v.31, n.2, p.152163, 2001.

DEBENAY, J.P., GUILLOU, J.J., REDOIS, F., GESLIN, E. Distribution trends of foraminiferal assemblages in paralic environments: a basis for using foraminifera as early warning indicators of anthropic stress. In: MARTIN, R. (ed.), **Environmental Micropaleontology:** Plenum Publishing Corporation, 2000. p.39-67.

DEBENAY, J.P.; EICHLER, B.B.; DULEBA, W.; BONETTI, C.; EICHLER-COELHO, P. Water stratification in coastal lagoons: its influence on foraminiferal assemblages in two Brazilian lagoons. **Marine Micropaleontology,** v.35, p.67-98, 1998.

DEBENAY J.P. Environmental monitoring through bioindicators: a French-Brazilian cooperation in the study of foraminifera. **Revista França - Flash Meio Ambiente,** v.9, p.1-4, 1996.

DEBENAY, J.P. Recent foraminiferal assemblages and their distribution related to environmental stress in the paralic environments of West Africa (Cape Timiris to Ebrie Lagoon).**The Journal of Foraminiferal Research.,** v.20, n.3, p.267-282, 1990.

DULEBA, W. Foraminifera and tecamebas as bioindicators of the hydrodynamic circulation of the Verde River estuary and Itacolomi Lake. In: MARQUES, O.A.V.; DULEBA, W.(Orgs.). **Juréia-Itatins ecological station:** physical environment, flora and fauna. led. Ribeirao Preto: Holos, 2004. v.1, 86-102p.

DULEBA, W.; DEBENAY, J. Pierre. Hydrodynamic circulation in the estuaries of Estaçao Ecológica Juréia-Itatins, Brazil, inferred from foraminifera and thecamoebian assemblages. **The Journal of Foraminiferal Research,** v.33, n.1, p.62-94, 2003.

DULEBA, W. Paleoenvironmental interpretations obtained from variations in the coloration of foraminifer shells at Enseada do Flamengo, SP. **Boletim do Instituto Oceanogràfico,** v.42, n.1-2, p.63-72, 1994.

DIAS-BRITO, D.; MOURA, J.A.; WÜRDIG, N. Relationships between ecological models based on Ostracodes and Foraminifers from Sepetiba Bay (Rio de Janeiro, Brazil). In: HANAY, T.; IKEYA, N.; ISHIZAKI, K (Eds.) **Evolutionary Biology of Ostracoda**, [s.l.]: Elsevier, 1988. 467-484p.

DOMINGUEZ, J.M.L. **Quaternary evolution of the coastal plain associated with the mouth of the Jequitinhonha River (BA): influence of sea level variations and coastal sediment drift.** Bahia. 1983. 79f. Dissertation (Master's Degree in Geology) - Institute of Geosciences, Federal University of Bahia, Salvador, 1983.

ESCOBAR, N.F.C. **Geochemistry of surface water and bottom sediment in the lower reaches of the Una, Pardo and Jequitinhonha rivers, Southern Bahia, Brazil.** 2013. 121f. Dissertation (Master's Degree) - Institute of Geosciences, Federal University of Bahia, Salvador, 2013.

EICHLER, P.P.B.; EICHLER, B.B.; DAVID, C.J.; MIRANDA, L.B.; SOUZA, E. The estuary ecosystem of Bertioga, Sao Paulo, Brazil. **Journal of Coastal Research,** v.2, p.1110-1113, 2006.

EICHLER, P.P.B. **Evaluation and diagnosis of the Bertioga Channel (Sao Paulo, Brazil) through the use of foraminifera as environmental indicators.** 2001. 240f Thesis (Doctorate in Oceanography) - Institute of Biological Oceanography, University of Sao Paulo, Sao Paulo, 2001.

EMBRAPA - Brazilian Agricultural Research Corporation. **Manual of soil analysis methods.** 2ed. Rio de Janeiro: National Soil Research Center, 1997. 212p.

EICHLER, B.B.; BONETTI, C. Distribution of foraminifera and tecamebas occurring in the Baguaçu River mangrove, Cananéia, Sao Paulo - Relations with environmental parameters.

Revista Pesquisas em Geociências, v.22, n.1-2, p.32-37, dec. 1995.

FERNANDES, S.; PILLAY, S. A study of the net flux of nitrates from estuaries of the eThekwini Municipality of Durban, KwaZulu-Natal, South Africa. **Environ Earth Sci.,** v.67, p.2193-2203, apr. 2012.

FRONTALINI, F.; COCCIONI, R. Benthic foraminifera for heavy metal pollution monitoring: A case study from the central Adriatic Sea coast of Italy. **Estuarine, Coastal and Shelf Science,** v.76, p.404-417, 2008.

FRONTALINI, F.; COCCIONI, R. Benthic foraminiferal for heavy metal pollution monitoring: A case study from the central Adriatic Sea coast of Italy. **Estuarine Coastal and Shelf Science,** v.76, n.2, p.404-417, jan. 2008.

FARIAS, L.G.Q. The challenge of sustainability in the coastal areas of southern Bahia. **Multidisciplinary Academic Journal,** n.12, p.1-10, 2007.

FERREIRA, V.O. **Paisagem, Recursos Hidricos e Desenvolvimento Econòmico na Bacia do Rio Jequitinhonha, MG.** 2007. 291f. Thesis (Doctorate in Geography) - Federal University of Minas Gerais, Belo Horizonte, 2007.

FONTANIER, C.; JORISSEN, F.J.; LICARI, L.; ALEXANDRE, A.; ANSCHUTZ, P.; CARBONEL, P. Live benthic foraminferal faunas from the Bay of Biscay: faunal density, coomposition, and microhabitat. **Deep-Sea Research,** v.49, n.4, p.751-785, apr. 2002.

FERNANDES, J.A. E PERIA, L.C.S. Caracteristicas do Ambiente. In: SHAEFFER- NOVELLI, Y. (Orgs.). **Mangroves:** ecosystems between land and sea. Sao Paulo: Caribbean Ecological Research, 1995. 3-16p.

GESLIN, E.; DEBENAY, J.P.; DULEBA, W.; BONETTI, C. Morphological abnormalities of foraminiferal tests in Brazilian environments: comparison between polluted and non-polluted areas. **Marine Micropaleontology,** v.45, n.2, p.151-168, jun. 2002.

HARRIS, L.A. AND BRUSH, M.J. Bridging the gap between empirical and mechanistic models of aquatic primary production with the metabolic theory of ecology: Na example from estuarine ecosystems. **Ecological Modelling,** v.233, p.83-89, may. 2012.

HICKMAN, C.S. AND LIPPS, J.H. Foraminiferivoy: Selective ingestion of foraminifera and rest alterations produced by the neogastropod Olivella. **The Journal of Foraminiferal Research,** v.13, p.108-114, 1983.

HABIB, B.R.; LARSON, B.F.; NUTTLE, W.K.; RIVERA-MONROY, V.H.; NELSON, B.R.; MESELHE, E.A.; TWILLEY, R.R. Effect of rainfall spatial variability and sampling on salinity prediction in an estuarine system. **Journal of Hydrology,** v.350, n.1, p.56-57, feb. 2008.

HOHENEGGER, J. The importance of symbiont-bearing benthic foraminifera for West Pacific carbonate beach environments. In BARBIERI, R., HOHENEGGER, J.; N. PUGLIESE (Eds):

Foraminifera and Environmental Micropaleontology, Environmental Micropaleontology Symposium at the 32nd International Geological Congress, 3. 2006. **Marine Micropaleontology,** v.61, n.1-3, p.4-39, 2006.

INMET, National Institute of Meteorology. 2012. **Meteorological charts.** Available at: <www.inmet.gov.br>. Accessed on: 07 Aug. 2012.

IMGA. Minas Gerais Institute of Water Management. **Monitoring the quality of surface water in the Jequitinhonha River Basin in 2009.** Belo Horizonte: Instituto Mineiro de Gestao das Aguas, 2010. 444p. (Annual report).

IBGE - Brazilian Institute of Geography and Statistics. **Environmental Diagnosis of the Jequitinhonha River Basin** - general guidelines for land use planning. Salvador, 1997. Available at: <http://biblioteca.ibge.gov.br>. Accessed on: April 19, 2014.

KOUKOUSIOURA, O.; DIMIZA, M.D.; TRIANTAPHYLLOU, M.V.; HALLOCK, P. Living benthic foraminifera as an environmental Proxy in coastal ecosystems: A casa study from the Aegean Sea (Greece, NE Mediterranean). **Journal of Marine Systems,** v.88, p.489-501, 2011.

KOHO, K. **Benthic foraminifera: ecological indicators of past and present oceanic environments - A glance at the modern assemblages from the Portuguese submarine canyons.** 2008. 166f. Thesis (Doctorate in Geology) - Utrecht University, Budapestiaan, 2008.

KATHIRESAN, K.; QASIM, S. Biodiversity of Mangrove Ecosystems. **Conservation and Society,** v.3, n.2, p.537-539, 2005.

LEMOS Jr., I.C. **Distribution and taphonomic aspects of recent foraminifera on the continental shelf of Sergipe, Brazil.** 2011. 100f. Dissertation (Master's in Geology) - Institute of Geosciences, Federal University of Bahia, Salvador, 2011.

LANÇONE, R.B.; DULEBA, W.; MAHIQUES, M. Bottom dynamics of Flamengo cove, Ubatuba, Brazil, inferred from the spatial distribution, morphometry and taphonomy of foraminifera. **Revista Brasileira de Paleontologia,** v.8, n.3, p.181-192, 2005.

LEITE, C.M.B.; BERNARDES, R.S.; SEBASTIÂO, A.O. Walkley-Black method for determining organic matter in soils contaminated by leachate. **Revista Brasileira de Engenharia Agricola e Ambiental,** v.8, n.1, p.111-115, 2004.

LEIPNITZ, I.I.; NOWATZKI, C.H.; LEIPNITZ, B.; AGUIAR, E.S.; OLIVEIRA, R.F.G.; GIOVANONI, L. Foraminifera from the Quaternary of Lagoa do Peixe, State of Rio Grande do Sul, Brazil. **Revista Brasileira de Paleontologia,** v.5, p.39-47, 2003.

LEVIN, L.A.; ETTER, R.J.; GOODAY, A.J.; SMITH, C.R.; PINEDA, J.; STUART, C.T.;

HESSLER, R.R.; PAWSON, D. Environmental influences on regional deep-sea species diversity. **Annual Review of Ecology, Evolution and Systematics,** v.32, p.51-93, 2001.

LACERDA, L.D. Trace Metals Biogeochemistry and Diffuse Pollution in Mangrove Ecosystems. **Mangrove Ecosystems Occasional Papers,** v.2, p.65-72, 1998.

LACERDA, L.D., CARVALHO, C.E.V., TANIZAKI, K.F., OVALLE, A.R.C.; REZENDE, C.E. The biogeochemistry and trace metals distribution of mangrove rhizospheres. **Biotropica,** v.25, p.252-257, 1993.

LEÂO Z.M.A.N.; MACHADO A.J. Variaçao da Cor dos Graos Carbonâticos de Sedimentos Marinhos Atuais. **Revista Brasileira de Geologia,** v.1, p.87-91, 1989.

LOEBLICH, A.R.; TAPPAN, H. Protista 2: Sarcodina chiefly "Thecamoebians" and foraminiferida. In: MOORE, R.C. (Ed.) **Treatise on Invertebrate Paleontology.** Meriden, The Meriden Gravure Company, 1978. v.1, p.55-139.

MOREIRA, I.T.A.; OLIVEIRA, O.M.C.; TRIGUIS, J.A.; QUEIROZ, A.F.S.; FERREIRA, S. L.C.; MARTINS, C.M.S; SILVA, A.C.M.; FALCÂO, B.A. Phytoremediation in mangrove sediments impacted by persistent total petroleum hydrocarbons (TPH's) using *Avicennia schaueriana.* **Marine Pollution Bulletin,** v.67, p.130-136, 2013.

MULLER, F.L.L.; TANKÉRÉ-MULLER, S.P.C. Seasonal variations in surface water chemistry at disturbed and pristine peatland sites in the Flow Country of northern Scotland. **Science of the Total Environment,** v.435-436, p.351-362, oct. 2012.

MAHER, D.; EYRE, B.D. Insights into estuarine benthic dissolved organic carbon (DOC) dynamics using δ 13C-DOC values, phospholipid fatty acids and dissolved organic nutrient fluxes. **Geochimica et Cosmochimica Acta,** v.75, n.7, p.1889-1902, 2011.

MORAES, S.S. **Spatial distribution and taphonomy of foraminifera on the continental shelf of the northern region of the dendê coast (mouth of the jequiriçà river to ponta dos castelhanos), State of Bahia.** 2006. 176f. Thesis (Doctorate in Geology) - Institute of Geosciences, Federal University of Bahia, Salvador, 2006.

MORIGI, C. Benthic environmental changes in the Eastern Mediterranean Sea during sapropel S5 deposition. **Palaeogeography, Palaeoclimatology, Palaeoecology,** v.273, n.23-4, p.258271, 2009.

MORAES, S.S.; MACHADO, A.J. Evaluation of the hydrodynamic conditions of two coastal reefs on the North Coast of the State of Bahia. **Revista Brasileira de Geociências,** v.33, n.2, p.201-210, 2003.

MIRANDA, L.B.; CASTRO, B.M.; KJERFVE, B. **Principios de oceanografia fisica de estuarios.**

Sao Paulo. EDUSP, 2002. 408p.

MORAES, S.S. **Interpretations of hydrodynamics and types of transport based on the study of recent foraminifera from the coastal reefs of Praia do Forte and Itacimirim, north coast of the State of Bahia.** 2001. 98f. Dissertation (Master's Degree in Geology) - Institute of Geosciences, Federal University of Bahia, 2001.

MACHADO, A.J.; SILVA, S.S.F.; BRAGA, Y.S.; MORAES, S.S.; NASCIMENTO, H.A.; MACÊDO, C.F.C.M. Genera of foraminifera from the reef area of Praia do Forte - North Coast of the State of Bahia. In: CUSHMAN FOUNDATION RESEARCH SYMPOSIUM, 7, 1999, Porto Seguro. Proceedings: Porto Seguro, Brazilian Association of Quaternary Studies. 7 Associaçao Brasileira de Estudos do Quaternàrio, 1999.

MACHADO A.J. Pyritized foraminifera from Iguape Bay, Bahia. **Acta Geol. Leopol.,** v.45, p.77-85, 1997.

MURRAY, J.W. **Ecology and Palaecology of Benthic Foraminifera.** New York: Longman. Scientific and Technical, 1991. 397p.

MURRAY, J.W.; WRIGHT, C.A. Surface textures of calcareous foraminiferids. **Palaeontology,** v.13, p.184-187, 1970.

MAIKLEM, W.R. Black and Brown speckled foraminiferal sand from the southern part of the great Barrier. **Reef Journal of Sedimentary Petrology,** v.37, p.1023-1030, 1967.

PRAZERES, M. F. **Taxocenosis of foraminifera and application of the FORAM index in the reef ecosystems of Abrolhos and Corumbau (BA).** 2007. 70f. Monograph (Graduation in Marine Biology) - Universidade Federal Fluminense, Niterói, 2007.

PRÓSPERI, V.A.; NASCIMENTO, I.A. Ecotoxicological assessment of marine and estuarine environments. In: ZAGATTO, P.A.; BERTOLETTI, E. (Eds.). **Aquatic ecotoxicology:** principles and applications sao carlos, sp. Sao Paulo: RIMA Editora, 2006. p.269-346.

PERH-BA. State Water Resources Plan for the State of Bahia. **Diagnosis and Regionalization.** Salvador, 2003 (Final report of stage I). Available at: <http://biblioteca.inga.ba.gov.br>. Accessed on: August 18, 2011.

QUEIROZ, A.F.S.; OLIVEIRA, O.M.C. **Geoenvironmental diagnosis of mangrove areas and development of technological processes applicable to the remediation of these areas:** subsidies for an impact prevention program in areas with potential for oil activities in the southern coastal region of the State of Bahia (PETROTECMANGUE-BASUL). Salvador: EDUFBA, 2013. 148p. (Technical report).

QUEIROZ, J.F.; TRIVINHO-STRIXINO, S.E.; NASCIMENTO, V.M.C. Organismos bentônicos bioindicadores da qualidade de água da bacia do mèdio Sao Francisco. **Série Comunicado Técnico da Embrapa Meio Ambiente**, v.3, p.1-4, 2000.

RODRIGUES, E.S.; UMBUZEIRO, G.A. Integrating toxicity testing in the wastewater management of chemical storage terminals - A proposal based on a ten-year study. **Journal of Hazardous Materials,** v.186, n.2, p.1909-1915, 2011.

REBOTIM, A.S. **Planktonic foraminifera as indicators of water bodies north and south of the Azores front/current: Evidence from abundance and stable isotope data.** 2009, 104f. Dissertation (Ecology in masters) - University of Porto, Porto, 2009.

ROSOT, M.A.; BARCZAK, C.L.; COSTA, D.M.B. Vulnerability analysis of the Itacorubi mangrove swamp: anthropic areas using satellite images and geoprocessing techniques. Florianópolis. In: Latin American Scientific Initiation Meeting, 14. 2011, Paraiba. **Electronic proceedings...** Vale do Paraiba: Universidade do Vale do Paraiba, 2000. p.1-6. Available at: <http://www.inicepg.univap.br/cd/INIC_2011/anais/arquivos/RE_0772_0642_01.pdf>.

Accessed on: March 24, 2013.

SRHSH - Secretariat of Water Resources, Sanitation and Housing. Regional Diagnosis, Characterization of the Physical Environment. **Superintendência de Recursos Hidricos (SRH),** v.1, p.194, 1996.

SANTOS, L.O. **Availability of chemical elements in apicuns and in sediments and leaves of mangroves in the Municipality of Madre de Deus, Bahia, Brazil.** 2013. 96f. Dissertation (Master's Degree in Geochemistry) - Institute of Geosciences, Federal University of Bahia, Salvador, 2013.

SODRÉ, F.; SCHNITZLER, D.; SCHEFFER, E.; GRASSI, M. Evaluating Copper Behavior in Urban Surface Waters Under Anthropic Influence. A Case Study from the Iguaçu River, Brazil. **Aquatic Geochemistry,** v.18, n.5, p.389-405, 2012.

SALEM, E.M.; MERCER, E.D. The Economic Value of Mangroves: A Meta-Analysis. **Sustainability,** v.4, p.359-383, 2012.

SARASWAT, R.; NIGAM, R.; PACHKHANDE, S. Difference in optimum temperature for growth and reproduction in benthic foraminifer Rosalina globularis: Implications for paleoclimatic studies. **Journal of Experimental Marine Bioloogy and Ecology,** v.405, n.1, p.105-110, 2011.

SEMENSATTO JR., D. L.; FUNO, R. H. F.; DIAS-BRITO, D.; COELHO JR., C. Foraminiferal ecological zonation along a Brazilian mangrove transect: diversity, morphotypes and the influence of subaereal exposure time. **Revue de Micropaléontologie,** v.52, p.67-74, 2009.

SANTOS, A.P. **Apropriação da Natureza e Produção do Espaço no Municipio de Belmonte - Bahia.** 2008.   146f. Dissertation (Master's Degree in Geography) - Institute of Geosciences, Federal University of Bahia, Salvador, 2007.

SANTA-CRUZ, J. **Current foraminifera in a mangrove impacted by oil 20 years ago: The Iriri river, Bertioga channel, Santos-SP.** 2004. 142f. Dissertation (Master's Degree in Geosciences and Environment) - Institute of Geosciences and Exact Sciences, São Paulo State University, São Paulo, 2004.

SEMENSATTO, J.R.; DIAS-BRITO, D. Preliminary environmental analysis of a tropical paralic area of the Sao Fransisco River Delta, Sergipe-Brazil, based on the synecology of foraminifera and tecamebas. **Revista Brasileira de Paleontologia,** v.7, n.1, p.53-66, 2004.

SANTOS, P.S., MARQUES, A.C.; ARAUJO, A. Remnants of coastal vegetation in the southeast region of Bahia - municipalities of Una and Canavieiras. In: Mostra de Talento Cientifico, 1., 2002, Curitiba. **Proceedings...** Curitiba: GIS Brasil, 2002. p.1-6.

SCHAEFFER-NOVELLI, Y.; CINTRÓN-MOLERO, G.; SOARES, M.L.; DE-ROSA, M.T. Brazilian mangroves. **Aquatic Ecosystem Health and Management,** v.3, p.561-570, 2000.

SAMIR A.M. The response of Benthic foraminifera and ostracods to various pollution sources: a study from two lagoons in Egypt. **Journal of Foraminiferal Research,** v.30, n.2, p.83-98, 2000.

SIMÕES, M. G.; HOLZ, M. Taphonomy: Fossilization processes and environments. In: CARVALHO, I.S (Org.). **Paleontologia.** 2ed. Rio de Janeiro: Interciência, p.98, 2000.

SCHAEFFER-NOVELLI, Y. **Mangrove:** ecosystem between land and sea. Sao Paulo: Caribbean Ecological Research, 1995. p.100.

SCHAEFFER-NOVELLI, Y.; CINTRON, G.; ADAIME, R. R.; CAMARGO, T. M. Variability of the mangrove ecosystem along the Brazilian Coast. **Estuaries,** v.13, n.2, p.204-218, 1980.

SWINCHATT, J.P. Significance of constituent composition, texture, and skeletal breakdown in some recent carbonates sediments. **Journal of sedimentary petrology,** v.35, n.1, p.71-90, 1965.

TEODORO, A.C.; DULEBA, W.; GUBITOSO, S.; PRADA, S.M.; LAMPARELLI, C.C.; BEVILACQUA, J.E. Analysis of foraminifera assemblages and sediment geochemical properties to characterize the environment near Araçà and Saco da Capela domestic sewage submarine outfalls of Sao Sebastiao Channel, Sao Paulo, Brazil. **Marine Pollution Bulletin,** v.60, n.4, p.536-553, 2010.

TOLER. S.K.; HALLOCK, P. Shell malformation in stressed Amphistegina populations: relation to biomineralization and paleoenvironmental potential. **Marine Micropaleontology,** v.34, p.107-115, 1998.

UEHARA-PRADO, M.; BROWN Jr., K.S.; FREITAS, A.V.L. Species richness, composition and abundance of fruit-feeding butterflies in the Brazilian Atlantic Forest: comparison between a fragmented and a continuous landscape. **Global Ecology and Biogeography,** v.16, p.43-54, 2007.

VASCONCELOS, A.O.; CELINO, J.J. Geology, geomorphology and evolution of coastal environments in the municipalities of Una, Canavieiras and Belmonte. In: CELINO, J.J.; HADLICH, G.M.; QUEIROZ, A.F.S.; OLIVEIRA, O.M.C. (Orgs). **Evaluation of coastal environments in the southern region of Bahia:** geochemistry, oil and society, Salvador: EDUFBA, 2014, 32p. 1 map, color. Scale 1:50.000.

VILELA, C.G. Foraminifera. In: CARVALHO, I.S. (Ed). **Paleontology.** 2ed. Rio de Janeiro: Interciência, 2004. 269-296p.

WETMORE, K.L. Correlations between test strength, morphology and habitat in some benthic foraminifera from the Coast of Washigton. **The Journal of Foraminiferal Research,** v.17, n.1, p.1-13, 1987.

WANILSON, L.S.; MATOS, R.H.R., KRISTOSCH, G.C. Geochemistry and geoaccumulation index of mercury in surface sediments of the Santos - Cubatao estuary (SP). **Quimica Nova**, v.25, n.5, p.753-756, 2005.

XAVIER, A.L.S. **Paleotectonics of the provenance areas and Petrography of the Salobro Formation, Pardo River Basin - Bahia**. 2009. 89f. Monograph (Graduation in Geology) - Institute of Geosciences, Federal University of Bahia, Salvador, 2009.

YANKO, V.; KRONFELD, J.; FLEXER, A. Response of benthic foraminifera to various pollution sources: implications for pollution monitoring. **Journal of Foraminiferal Research,** v.24, p.1-17, 1994.

ZILI, L.; ZAGHBIB-TURKI, D.; ARENILLAS, I. Biostratigrafía y eventoestratigrafía com foraminiferos plactónicos a través Del limite Paleoceno Eoceno en Kharrouba (Tùnez). **Geo-Themes,** v.10, p.1286-1288, 2008.

ZUCON, M.H.; LOYOLA, E.; SILVA, J. Distribuiçao de Foraminiferos e Tecamebas do Estuàrio do Rio Piaui, Sergipe. **Neuritica,** v.7, n.1-2, p.57-69, 1992.

ZANINETTI, L.; BRO'NNIMANN, P.; DIAS-BRITO, D.; ARAI, M.; CASALETTI, P.; KOUTSOUKOS, E.; SILVEIRA, S. Distribution e'cologique des foraminife'res dans la mangrove d'Acupe, Etat de Bahia, Bre'sil. Notes Lab. Paleontol. **Université de Genève,** v.4, n.1, p.1-17, 1979.

APPENDIX A - GEOGRAPHICAL COORDINATES

Table 1 - Geographical coordinates in UTM of the collection points in the estuarine channels

UTM SAMPLES N		UTM S
Una River		
UNA01	8.313.959	500.219
UNA02	8.314.992	500.024
UNA03	8.312.217	499.468
UNA04	8.310.885	499.189
UNA05	8.310.281	499.240
UNA06	8.308.590	499.180
UNA07	8.309.473	498.442
UNA08	8.308.608	497.021
UNA09	8.307.516	496.448
UNA10	8.308.406	495.638
Rio Pardo		
PDO01	8.313.959	500.219
PDO02	8.314.992	500.024
PDO03	8.312.217	499.468
PDO04	8.310.885	499.189
PDO05	8.310.281	499.240
PDO06	8.308.590	499.180
PDO07	8.309.473	498.442
PDO08	8.308.608	497.021
PDO09	8.307.516	496.448
PDO10	8.308.406	495.638
Jequitinhonha River		
JEQ01	8.249.534	515.187
JEQ02	8.248.633	514.358
JEQ03	8.247.425	515.151
JEQ04	8.247.507	514.529
JEQ05	8.247.657	513.820
JEQ06	8.247.586	512.841
JEQ07	8.247.141	512.866
JEQ08	8.247.253	512.352
JEQ09	8.246.307	511.207
JEQ10	8.246.198	510.314

Table 2 - Geographical coordinates in UTM of the collection points in the mangrove zones of the estuarine channels

SAMPLES	UTM N	UTM S
Mangrove area (Una River)		
UNA01	8.315.307	500.002
UNA02	8.314.314	500.315
UNA03	8.313.336	500.206
UNA04	8.312.963	500.159
UNA05	8.308.649	498.738
UNA06	8.311.551	499.612
Mangrove area (Pardo River)		

PDO01	8.263.608	505.300
PDO02	8.264.610	505.246
PDO03	8.261.620	505.301
PDO04	8.265.619	505.315
PDO05	8.265.371	504.487
PDO06	8.266.520	503.769
Mangrove area (Jequitinhonha River)		
JEQ01	8.249.718	515.153
JEQ02	8.248.989	515.324
JEQ03	8.249.517	514.889
JEQ04	8.248.353	514.544
JEQ05	8.248.586	514.326
JEQ06	8.248.896	513.873

APPENDIX B - FORAMINIFERA FAUNA

Taxonomy

The identification of foraminifera at species level was based on existing literature by various authors, but the systematic classification was based on Loeblich and Tappan (1988). 13 benthic and 2 planktonic species were obtained, distributed in 16 genera and 7 suborders, namely:

PROCTIST Kingdom Haeckel, 1866

Phylum GRANULORETICULOSA Margulis, 1999

Class FORAMINIFERIDA Sen Gupta, 1999

LITUOLIDA Order of Blainville, 1827

Superfamily LITUOLACEA de Blainville, 1827

Family HAPLOPHRAGMOIDIDAE Maync, 1952

Genus *Haplophragmoides* Cushman, 1910

Haplophragmoides wilberti Andersen, 1953

Order TROCHAMMININA Saidova, 1981

Superfamily TROCHAMMINACEA Schwager, 1877

Family TROCHAMMINIDAE Schwager, 1877

Subfamily TROCHAMMININAE, Schwager, 1877

Genus *Trochammina,* Parker & Jones, 1859

Trochammina inflata, Montagu, 1808

Trochammina squamata, Jones & Parker, 1860

TEXTULARIIDA Order Delage and Hérouard, 1896

Superfamily TEXTULARIACEA Ehrenberg, 1838

Family TEXTULARIIDAE Ehrenberg, 1838

Subfamily TEXTULARIINAE Ehrenberg, 1838

Genus *Textularia* Defrance, 1824

Textularia agglutinans (d'Orbigny, 1839)

Order MILIOLIDA Delage & Hérouard, 1896

Family MILIAMMINIDAE Saidova, 1981

Genus *Miliammina* Heron-Allen & Earland, 1930

Miliammina fusca Brady, 1870

Superfamily MILIOLACEA Ehrenberg, 1839

Family HAUERINIDAE Schwager, 1876

Subfamily MILIOLINELLINAE Vella, 1957

Genus *Pyrgo* Defrance, 1824

Pyrgo nasuta Cushman, 1935

Genus *Triloculina* d'Orbigny, 1826

Triloculina sp.

Subfamily HAUERININAE Schwager, 1876

Genus *Quinqueloculina* d'Orbigny, 1826

Quinqueloculina fusca Brady, 1870

Quinqueloculina lamarckiana d'Orbigny, 1839

Quinqueloculina seminula Linnaeus, 1758

Quinqueloculina venusta Karrer, 1868

Order SPIRILLINIDA Hohenegger & Piller, 1975

Family SPIRILLINIDAE, Reuss & Fritsch, 1861

Genus *Spirillina* Ehrenberg 1843

Spirillina decorata Brady, 1884

Spirillina sp.

Order ROTALIIDA Delage and Hérouard, 1896

Superfamily ROTALIACEA Ehrenberg, 1839

Family ROTALIIDAE Enrenberg, 1839

Subfamily AMMONIINAE Saidova, 1981

Genus *Ammonia* Brünnich, 1772

Ammonia beccarii (Linnaeus, 1758)

Ammonia Parkensoniana (d'Orbigny, 1839)

Ammonia tepida (Cushman, 1926)

Family ELPHIDIIDAE Galloway, 1933

Subfamily ELPHIDIUM bartletti Cushman, 1933

Genus *Elphidium* de Montfort, 1808

Elphidium bartletti Cushman, 1933

Elphidium discoidale (d'Orbigny, 1839)

Elphidium excavatum (Terquem, 1875)

Elphidium poeyanum (d'Orbigny, 1826)

Elphidium sagrum (d'Orbigny, 1839)

Superfamily NONIONACEA Schultze, 1854

Family NONIONIDAE Schultze, 1854

Subfamily NONIONINAE Schultze, 1854

Genus *Nonion* de Montfort, 1808

Nonion grateloupi d'Orbigny, 1826

Superfamily PLANORBULINACEA Schwager, 1877

Family CIBICIDIDAE Cushman, 1927

Subfamily CIBICIDINAE Cushman, 1927

Genus *Cibicides* de Montfort, 1808

Cibicides sp.

Order GLOBIGERINIDA Delage and Hérouard, 1896

Superfamily GLOBIGERINACEA Carpenter, Parker and Jones, 1862

Family GLOBIGERINIDAE Carpenter, Parker and Jones, 1862

Subfamily GLOBIGERININAE Carpenter, Parker and Jones, 1862

Genus *Globigerina* d'Orbigny, 1826

Globigerina pachyderma (Ehrenberg, 1861)

Globigerina trilobus d'Orbigny, 1826

116

Order LAGENIDA Delage and Hérouard, 1896

Superfamily BOLIVINACEA Glaessner, 1937

Family BOLIVINIDAE Glaessner, 1937

Genus *Bolivina* d'Orbigny, 1839

Bolivina difformis Williamson, 1858

Bolivina laevigata Karrer, 1878

Bolivina pulchella (d'Orbigny, 1839)

Bolivina striatula Cushman, 1922

Genus *Brizalina* Costa, 1856

Brizalina alata Seguenza, 1862

Order ASTRORHIZIDA, Lankester, 1885

Superfamily SACCAMMINACEA Brady, 1884

Family SACCAMMINIDAE Brady, 1884

Subfamily SACCAMMININAE Brady, 1884

Genus *Saccammina* Sars, 1869

Saccammina sphaerica Brady, 1871